As If They Were Ours

The Story of Camp Tyson
America's Only
Barrage Balloon Training Facility

Shannon McFarlin

MERRIAM PRESS

WORLD WAR II HISTORY 7
BENNINGTON, VERMONT
2016

ISBN 9781576384732
Library of Congress Control Number: 2015959957
Merriam Press #WH7-P

This work was designed, produced, and published in
the United States of America by the

Merriam Press
133 Elm Street Suite 3R
Bennington VT 05201

E-mail: ray@merriam-press.com
Web site: merriam-press.com

The Merriam Press publishes new manuscripts on historical subjects, especially military history and with an emphasis on World War II, as well as reprinting previously published works, including reports, documents, manuals, articles and other materials on historical topics.

Dedication

To my Great-Grandmother, Lillie Foy Brady
my Grandmother, Ellie Brady Snow
and my Mother, Jeannette Snow McFarlin

whose stories about Camp Tyson inspired me to write this book.

Acknowledgements

WHERE to begin? There are so many people whose help was invaluable in producing this book. I've known since I was young that I wanted to write a book about Camp Tyson and have been collecting stories and insights from myriad people for years.

My family, of course, was the main inspiration. My family was just one of many in Paris and Henry County, Tennessee, which housed construction workers and soldiers and their families from Camp Tyson during World War II. I began asking questions about the camp when I was a little girl and had found a photo of an unknown man and woman in one of our family scrapbooks. I learned they were a soldier and his wife who lived in a small room on our back porch and that they had become part of the family during that time.

More questions and more information: the soldier and his family weren't the only ones who lived with us during World War II. At one point, I learned, we had five families living in our tiny home, including the families of construction workers who had helped build the camp. How all those people were able to fit in our small house, maintain each family's privacy and arrange everyone's schedules of meals, bathroom time, and relaxation is still puzzling to me.

In Paris, there are reminders of Camp Tyson all around, if you know where to look: the rows and rows of identical, compact brick 'Camp Tyson' or 'government' houses that were built during the war to keep up with the population boom that occurred; Tyson Avenue, one of the main thoroughfares in and out of Paris, which originally was called the Camp Tyson Highway and which linked Paris directly to the camp; and the mortars and ordnance which periodically are still found buried in lawns in the neighborhoods where soldiers' families used to live.

And driving a few miles outside of Paris (on the old Tyson highway, naturally), you arrive at the small town of Routon and the former Camp itself. Now operated as a clay mining company, the sign in front of the facility reminds passersby of its heritage by noting it is the Lhoist Spinks Clay Co., Camp Tyson plant.

Remnants of Camp Tyson are still evident within the clay company grounds. The 90' hangar in which barrage balloons were built is still there, as are the old incinerator, motor pool buildings, office buildings, and miles of old roadways, now infiltrated with weeds, and which still are lined with the foundations of old barracks in the ground. Looking at the still-existing foundations and ornamental trees and flowers, you can still visualize just where officers' housing, barracks and the hospital buildings were situated. The home in which General John Maynard and his family lived is still on the grounds, and you can see where the main guard booth was located on Gate 3 Rd.

Our family used to drive through the grounds periodically, as I'm sure other local families did, just to visualize how the camp used to look and relive its glory days. Those drives and the conversations they produced also were a source of inspiration.

Many years have passed since the end of World War II, and with that passage of time, memories have faded and sadly, many of those associated with the camp, including most of the Camp Tyson soldiers, are now deceased. Even more sadly for me, several of the people who I interviewed — and whom had told me how excited they were that a book was being written — passed away over the course of the year and 8 months it took for me to write the book.

Compiling the information for the book may have seemed like a daunting task, but thankfully the project was eased by so many people who were so generous with their time, memories and memorabilia. One of the best surprises of all was discovering just how many people in Paris and the surrounding area have kept their Camp Tyson mementoes. Many, many collections of memorabilia exist in local households, which I recognized as a testament to just how important Camp Tyson was to Henry County families.

First and foremost, I thank the Tennessee River Resort Act (TRRA) board of directors for funding the project. Former Henry County Alliance CEO Carl Holder, a history lover himself, recognized early the importance of preserving the Camp Tyson story and enthusiastically endorsed the project. The TRRA board members likewise were enthusiastic in their support, voting unanimously to finance the project, and for that I am grateful.

Special thanks to those whose help was invaluable:

Stephanie Routon Tayloe, Henry County Archivist, who generously gave of her time and knowledge of local history.

Randy Ford, assistant to former Congressman John Tanner, who navigated my Freedom of Information Act (FOIA) request through the bureaucratic red tape to search U.S. Army and government files for material. With his help, I was able to access the Wiley Report, a secret document which the U.S. Army commissioned to produce information on black soldiers, their training, their activities and racial unrest at several Army camps.

John McDougal of The Paris Post-Intelligencer, who searched the newspaper's archives for pertinent articles for me on numerous occasions.

Susan Gordon of The Tennessee Historical Society, whose videotaped interviews in Paris with local soldiers, their wives, and former employees at the camp in April of 1992, was a priceless source of information, especially since several of those she interviewed were deceased when I began my project. Those interviews were conducted for the Historical Society's "Home Front" project.

Laura Lohr of Waterfield Library at Murray State University, who performed numerous interlibrary loan requests for me.

Bill Davison of Waynesburg, Pennsylvania, who I came to know early in the project. Davison's father was a member of the 320th Anti-Aircraft Barrage Balloon Battalion, the only all-black unit to serve on D-Day. Davison has for years been encouraging the government and national organizations to recognize the importance of the 320th.

Former Paris Parks and Recreation Department employee Don Williams, who found a notation in an old city cemetery log that a soldier from the 320th Battalion was listed as being buried in an unmarked grave at Maplewood Cemetery in Paris. Using his 'water witching' method, Williams also was able to confirm for me that a casket was indeed buried at the grave, clearing the way for me to further investigate the soldier's identity and the situation surrounding his death.

John Gartrell of The Afro-American of Baltimore, who opened the archives of the newspaper to me.

Linda Hervieux of Paris, France, a writer with The New York Daily News, who searched the archives at the National Archives and Records Administration (NARA) for pertinent material while on a trip to Washington, D.C. Hervieux became interested in the 320th while doing a story on William Dabney's Medal of Honor award and had planned a research trip to the NARA. While there, she generously searched the archives for material on Camp Tyson, the 320th, and the prisoners of war who were housed there.

Dr. Bill Mulligan, history professor of Murray State University, who was my master's thesis advisor when I earned my master's degree in public history from Murray State. He has remained a friend and, with his background as a book author, has provided indispensable advice for this project.

Dr. Mulligan, Virgil and Jo Wall and Carl Holder for proof-reading the first drafts of the book. Susan Jones, Pat Terrell and Rick Owens for proof-reading later drafts of the book.

Several people whose collections of Camp Tyson memorabilia and photographs were a valuable source of information for me, including Lou and Bettye Carter, Bennye Phillips, Eddie Moody, Jeanne Townsend, Loyal Whiteside, James Wilson, Joe Lankford, Virgil and Jo Wall, Roland Parkhill, Red Boden, Jerry Ridgeway, Brenda Lewis, Nelda Pinson, Val Routon and Stephanie Routon Tayloe, and Wayne Webb.

A former soldier at the camp, James Wilson also was an invaluable help the day he and I drove around the grounds of the camp. His memory of where barracks, buildings, churches and the POW camp used to be situated was a huge help.

The late Rebecca Goins, who had kept a suitcase full of legal documents which her late husband had collected while embroiled with a dispute with the Spinks Clay Co. over access to the camp's grounds from Gate 3 Rd. Rebecca was a big help to me, as was Joe Hill, the conservator of her estate, who allowed me to use the suitcase after her death.

Dickie Carothers, for being generous with his time and knowledge of the acquisition of the Camp Tyson grounds by his family for the Spinks Clay Co. He also has a large collection of legal documents which he allowed me to photocopy and which were a big help in writing this book.

Chris Corley, Dan Collins, Frank Planchart, and all the other employees at Lhoist Spinks Clay Co., who have recognized the importance of preserving the history of the camp. Chris took me for a tour of the grounds of the former camp at the outset of my work on this book and was responsive to my various questions, Dan allowed me free access to boxes of photographs and materials that the Spinks company has kept on the history of the camp, and Frank also steered us around the grounds when James Wilson and I were having trouble finding General Maynard's old house.

AS IF THEY WERE OURS

Mary Will Gardner of Paris, who at the age of 100, was still sharper than many of us younger folk. Her memory helped me on more than one occasion as I was trying to reconstruct the local African-American experience during World War II.

I interviewed over 100 people for this book and all were giving of their time and insights. I thank them for sharing their memories with me and allowing me to share those memories with the readers of this book.

Contents

Preface

UNTIL now, Camp Tyson has only lived in the collective memories of its former soldiers and the people of Henry County, Tennessee, and the nearby area.

Regrettable, since the Camp Tyson experience was one of the most unique in Tennessee and United States history.

Camp Tyson was the only barrage balloon training facility in the United States during World War II, which meant the soldiers who served their shared activities, instruction and routines unlike any experienced by other servicemen during the war. Likewise, the people of Henry County were in the enviable position to witness the giant balloons flying aloft over the camp on a daily basis, a romantic and awe-inspiring sight for many residents. No other people in the country could boast proximity to such a phenomenon.

Of special significance is that Camp Tyson was home to the 320th Anti-Aircraft Barrage Balloon Battalion, the only all-African American unit to participate in the D-Day invasion. Utilizing the specialized training they received at the camp, the men of the 320th were the first in line off the aircraft carriers, launched from their positions by the balloons attached to their belts. Many did not survive and until the last few years, their achievements went unheralded and unrecognized.

It has only been recently — through newspapers articles on recent D-Day anniversaries — that the men were recognized for their achievements. Of all the men who served in the 320th, William Dabney has received the highest honor, being presented with the Legion of Honor in Paris, France, in 2009.

The 320th included a handful of men who became noted for other achievements, however, including the late Bill Pinkney, bass vocalist for The Drifters; George Dennis Leaks, who was an early member of the legendary Dixie Hummingbirds, and James Wilbert Pulley, a chauffeur to Baltimore politicians, including Mayor Thomas D'Alesandro and his young daughter, Nancy Pelosi, who became Speaker of the U.S. House of Representatives.

This book represents the first time the story of 320th has been told.

Camp Tyson and its magnificent flying barrage balloons may well have been an inspiration in the life of a young future astronaut. Gordon Cooper Sr. was a soldier stationed at the camp while his son was young. His son, Gordon Cooper Jr. became better known as "Gordo Cooper," one of the original Mercury 7 astronauts.

The Camp Tyson experience was unique other ways, too. Perhaps because it was a training facility, the officers and soldiers often brought their families with them. Henry County, like other Southern rural areas, was hit especially hard during the Depression, which meant housing was at a premium. The people of Paris, located a few miles from the camp, opened the doors of their homes to the soldiers and their families — as well as the hundreds of construction workers who came before them to build Camp Tyson.

Over the course of the camp's existence, most every house in Paris — including attics, back porches and sheds — were crowded with the newcomers, making the people of Paris an intrinsic part of the camp. Even over 70 years later, when asked about Camp Tyson, peoples' eyes light up with excitement, eager to share their memories. They know they were a part of something special.

Additionally, Camp Tyson was a Prisoner of War (POW) camp, housing thousands of German and Italian prisoners in the last years the camp was open, providing distinctive experiences for the soldiers and townsfolk with whom they came in contact.

It is a testament to how important the camp was to people that so many have kept their photographs and mementoes, from officers' chairs to paintings done by POWs to canteens and pieces of barrage balloons. Several even had written down their memories of the camp and kept them in notebooks that they saved and many people said they had saved their mementoes specifically hoping someone would come along to record the story.

And it wasn't just Henry Countians who preserved their recollections. From all across the country, soldiers wrote down their memoirs of the camp. Lengthy and detailed accounts of their training with the balloons, their memories of Paris and Henry County and how much a part of the local community they became while stationed there have been preserved, thanks to the internet.

Over 70 years past, you would expect memories to fade, but not so in the case of most who were interviewed for this book. Their reminiscences were as vivid as when they lived them; their excitement as palpable. With this book, the history of Camp Tyson is finally documented.

Chapter 1

Eddie Finds A New Home

EDDIE Clericuzio was in basic training in Fort Dix, N.Y., when he was notified he would be shipped out that day. But he wasn't told where he would be going.

He and his fellow soldiers were put on a train. Traveling through the days and nights — he couldn't remember exactly how many — they arrived at an out-of-the-way country outpost during a fierce rain storm in the middle of the night.

They disembarked the train and were ordered to walk through the mud a few miles to a just-constructed army barracks — way out in the middle of a field.

"There were no roads, no nothing," he said. And their Army superiors still hadn't told the soldiers where they were.

The barracks was just a shell, with no cots, chairs or tables. The soldiers were exhausted and just wanted to sleep, but they were ordered to get the barracks ready. "We swept up the floors which were covered in saw dust and we set up the cots and the furniture," Clericuzio said.

Then their superiors said they could make a phone call. Clericuzio, who had just been "living life," as he put it, when he was called to serve his country, called his mother back home in Bloomfield, N.J.

His mother asked him, "Where are you?"

Clericuzio said, "I don't know. I think I'm at the end of the world."

Well, he wasn't at the end of the world, but you couldn't blame him for thinking so.

He was now part of Camp Tyson, the United States' only barrage balloon training facility during World War II, which in its earliest stages was operated as a top secret from the public.

Over time, Clericuzio would change his first impression of his new home and its out-of-the-way location. So much so that he married a local girl and settled in nearby McKenzie, operating the theater there for decades.

As an Italian from the North, he had a lot to adjust to at Camp Tyson — the hot, humid weather in the summer, the strange accents of the Southerners, the food.

But over the years he was stationed there, he realized he had a front row seat to history. As a member of the first regiment to arrive at Camp Tyson, he watched the camp grow from a few barracks that had just been built by civilian construction workers to its eventual massive size.

From mud and farm fields, he saw the camp develop into a modern facility with 10 miles of asphalt roads, five miles of railroad, a hospital, movie theater, post office, industrial-sized laundry, two chapels, service club, several PX facilities, library, state-of-the-art sewer and electrical systems, motor pools, numerous offices, hydrogen generating plant, and its own water supply system which was symbolized by the huge water tower that seemed to stand sentry over the camp. And, of course, the 90' tall barrage balloon hangar which still is on the grounds.

Over time, the camp would grow to 400 buildings in all and by the end of World War II, it was home to 25,000 soldiers, including members of the Army Nurse Corps. Some 3,000 German and Italian prisoners of war were quartered there in the waning days of the war.

Camp Tyson was also home to the 320th Anti-Aircraft Barrage Balloon Battalion, the only all-black unit to land on the beaches of Normandy on D-Day June 6, 1944. Their contributions to the war effort were overlooked until recently when a handful began receiving recognition for their wartime bravery.

The highest honor among that group was bestowed upon William Dabney of Roanoke, Va., who was awarded The Legion of Honor from President Barack Obama in June 2009.

Locating the United States' only barrage balloon training facility in rural Henry County, Tennessee — which was especially rural and out-of-the-way in the 1940s — was no accident.

The Army had begun experimenting with barrage balloons as a defensive measure, even before Pearl Harbor was attacked in 1941. A temporary facility in Camp Davis, North Carolina, was established with some training maneuvers conducted there.

But the small community of Routon, Tennessee, was selected as the permanent facility in April of 1941. It was chosen for several reasons: it was out of the way of regular air traffic, it boasted natural and reliable wind currents that would be necessary for the balloon training, and it was located on a plentiful water supply.

Locating the facility in Henry County was a boon to the local area. Soon, hundreds of men and their families began streaming into Paris, looking for construction jobs and places to live. Traffic on the roads to Paris was bumper-to-bumper and it continued that way until the camp was opened, when soldiers also began knocking on doors, seeking homes for their families.

The city of Paris opened its arms to them, providing housing in every available nook and cranny, from sheds to back porches. Soon, most every home in the town was crowded, or should we say, over-crowded.

While history was being made at Camp Tyson and as it grew and grew, so did Paris. Camp Tyson impacted the city greatly, transforming it from a small town to a city.

Overnight, Paris became a boom town.

It is no exaggeration to say that most everyone in Henry County owed their livelihood to Camp Tyson. Rural West Tennessee was still reeling from the Depression. Camp Tyson produced much-needed jobs, both at the camp itself and in the inevitable side businesses that developed.

Quick-minded local entrepreneurs produced their own enterprises, from providing bus service to boarding facilities to sandwiches for troop trains. Even the youngsters got into it, with Coca-Cola machines and shoe-shining ventures.

Old-timers in Paris credit the camp with making the town a much richer place culturally from the influence that the soldiers brought — especially those who later settled here and brought their own heritage and cultures to the local area.

It is a testament to the impact that Camp Tyson had that so many people still remember their days working there or hosting soldiers and their families in such detail, even though so much time has passed. Most every household has kept some remembrance of the camp — whether it be postcards, phone books, or pillow cases.

A real bond developed because of Camp Tyson that endures today. Many local people are still in touch with the former soldiers and families who stayed with them over 60 years ago.

"They became like family" is the refrain heard over and over again.

It was a difficult time, but also an exciting time. Spurred by the anxiety of war, it was a time for sacrifice, hard work, fast changes, and fast friendships, high-flying balloons, and romance.

This is the story of Camp Tyson.

Chapter 2

Motors Roared, Hammers Thudded And Saws Swished

THE people of Henry County first received the news on August 15, 1941, that the U.S. government would purchase 2,000 acres of land near the little town of Routon, seven miles outside of Paris, to build an army camp.

To be housed at the camp was a barrage balloon training center — the only one of its kind in the U.S. No one in Henry County yet knew what a barrage balloon was or what purpose it would serve in the war effort. And frankly, they didn't care.

What they did care about was the announcement itself. Construction of an army camp meant one thing — jobs and lots of them — and in West Tennessee, hard-hit by the Depression, that was of utmost importance.

Contracts were quickly let, with Rock City-Strider Construction Co. in Nashville awarded the prime job as construction managers. The company established its headquarters in the O.C. Barton family mansion on N. Poplar St. in Paris, which today houses the Paris-Henry County Heritage Center.

"Building materials began to flow to the site by truck and trainload and the construction of Camp Tyson was underway," according to an article in the sesquicentennial edition of *The Paris Post-Intelligencer*.[1]

Hundreds of workmen "began pouring into the area and motors roared, hammers thudded and saws swished. Buildings began to shape along well-defined plans on attractively laid out streets hewn from open fields and woods."[2]

With Paris and Henry County "jammed to capacity with workmen," rapid progress was made in the construction. "Floodlights lighted the nights so that work might continue on a 24-hour basis, and Paris geared itself to a 24-hour day in order to care for the needs of thousands of men," according to the newspaper. Streams of automobiles

---

[1] "Paris Hummed 24 Hours A Day In Tyson Era," *The Paris Post-Intelligencer* Sesquicentennial Edition, Sept. 25, 1973, page 4C

[2] Ibid.

carrying thousands of people "formed a solid line along the seven miles between Paris and the camp site each morning at 6 o'clock and again each afternoon at four, when the day shift went off duty."[3]

At the peak of employment, some 8000 men worked on building the camp, with the peak reached around Christmas in 1941. In the ensuing months, the number of workmen decreased until the camp was completed on March 14, 1942.

Such an undertaking involved a great amount of work, much of which was performed before the first nail was hammered into the first piece of board.

Preparation for the project began with surveys and securing of property for the camp. Securing private property for a public project is always a controversial prospect and this was as controversial as any. It required acquisition of large amounts of acreage from numerous property owners and resulted in heartbreak for many whose homes and farmland were destroyed to make way for the camp.

Years later, Bryant Williams, whose family owned *The Paris Post-Intelligencer*, which is known to the locals as "The P-I," recalled the U.S. government exercised its right of eminent domain "and they bought farms whether the people wanted to sell or not. And a lot of them didn't."[4]

One of the first public domain acquisitions involved land which would be used to build a highway from Paris to the camp. Up to that point, a dirt back road — full of hills and curves — carried travelers bumpily back and forth along what is now called "The Old McKenzie Highway" and onto Dunlap Street in Paris.

With the increased traffic, that road was woefully inadequate. On December 12, 1941, five days after Pearl Harbor, Henry County Judge A.L. Paschall was notified that the Works Progress Administration (WPA) proposed to construct a highway to the Camp Tyson area.

The first six land owners along the new highway were sued for their property on December 19, 1941. Among them was the late Carlyle Hill, whose daughter Rebecca Goins, had kept all of the legal documents pertaining to the acquisition of that piece of property, along with more documents involving use of the Hill property for the camp itself.

The Hills were one of the property owners who were able to purchase their land back years later. Until her death on September 27,

---

[3]   Ibid.
[4]   Bryant Williams, personal interview with author, April 7, 2006.

2010, Goins lived on Gate 3 Rd., so-named when it was utilized for the camp. Goins' husband, Roy Goins accumulated all of the legal documents during a protracted battle he had with the Spinks' Clay Co., which purchased the Camp Tyson property after the war. At issue were right-of-way issues along Gate 3 Rd.

According to the records, the road was to cost $100,000, with the WPA paying 96 percent of the cost and Henry County paying for the right-of-way. The road was black-top and a county court jury approved the acquisition on May 7, 1942.[5]

Ironically, even with the Hill family's issues over the property acquisition, Carlyle Hill helped with construction of Camp Tyson. "Daddy believed that whatever needed to be done for peace should be done," Goins said.[6]

The project moved forward, with blueprints prepared for the buildings — all 400 of them — the camp layout itself, the camp's water and electric systems, and numerous roads. Clearing the land was a major operation. "It was necessary to remove all vestige of civilian habitation, including the wrecking of homes, barns, stables, cribs, pig pens, removal of fencing, filling in ponds and clearing away the brush growth in the 2,000 acre tract," according to *The Paris Post-Intelligencer*.[7]

As bulldozers and drag lines pushed trees and red clay into low spots, the outlines of a small city began to appear. The first roadways were solidly-packed clay and dirt to accommodate travel by the construction workers, but the roads soon were paved with asphalt as the project progressed.

Concrete foundations for the buildings were installed. The L&N Railroad tracks brought rail car-loads of lumber right to the site on its newly-completed railroad spur. A large sawmill was constructed on site to facilitate the construction.

"Trucks darted hither and yon bearing the much-needed supplies for the carpenters, bricklayers and other skilled laborers. Nothing was spared to speed up the operation, and no sooner was one phase of the work complete than another started," according to *The Paris Post-Intelligencer*. "Painters started wielding their brushes while carpenters

---

[5]   Henry County Circuit Court, docket 2309, May 7, 1942, property of Rebecca Goins.

[6]   Rebecca Goins, personal interview with author, Oct. 2, 2009.

[7]   *The Paris Post-Intelligencer*, sesquicentennial edition, page 4C.

were still working on other parts of the buildings. Heating equipment was installed before the painters had finished."[8]

"IT WAS FASCINATING TO SEE"

For a young newspaper reporter, Bryant Williams said, it was an exciting time. "I was out there quite a bit while it was being built. It was fascinating to see the paved streets, the sewer system installed, the beautiful new buildings, the barracks, the theater being built. It was a first-class operation."[9]

Williams recalled, "Workmen by the thousands showed up to help build it and they spent what money they had in Paris. It really helped our economy get back on its feet."[10] Williams was transferred overseas himself and did not return to Paris until 1946, after the camp was closed. Williams later became publisher of the family business, a position he performed for several years.

With the attack on Pearl Harbor on December 7, 1941, the war effort was accelerated. Soldiers began arriving at the camp even before it was completed. Eddie Clericuzio of McKenzie and Howard Koenen of Murray, Kentucky, were among soldiers who were greeted by incomplete barracks and sawdust upon their arrival.[11]

Many local and area companies benefited from the construction, by being awarded contracts. "Much of the building material was purchased in job lots direct from the jobber or manufacturer, but there was a very great amount of goods purchased from local merchants and businessmen," *The Paris Post-Intelligencer* pointed out in its sesquicentennial edition. "Not only did Paris benefit from the construction, but so did all neighboring towns."[12]

The city of Paris was "packed and jammed with workers, who slept in any kind of bed they could find. The shortage of living quarters was great, and rooms, apartments and homes were at a premium. Restaurants and lunch stands did a land office business and it seemed that all Paris flourished."[13]

---

[8]   Ibid.
[9]   Williams interview.
[10] Williams interview.
[11] Eddie Clericuzio and Howard Koenen, personal interviews.
[12] *The Post-Intelligencer*, sesquicentennial edition, page 4C.
[13] Ibid.

From 200 miles away, *The Memphis Commercial Appeal* also took note of the camp, with a full-page feature article complete with photographs, published Sunday, January 18, 1942. Reporter Bruce Tucker spent the day in Paris and at the camp and wrote with awe of the camp's construction.

Tucker noted he couldn't write as much about the camp as he could have before the Pearl Harbor attack "because there are a lot of things about our barrage balloons the enemy might like to know. The various things the men are being trained to do, and the things they're to do them with will be vague in the telling." [14]

Tucker noted that Paris was benefiting greatly from something it had made no effort to procure. "Peculiarly Paris made no effort to get the camp." A year before, Tucker had visited other cities that were angling for an army camp, such as Milan, Tennessee. During his tour of the area, he visited Paris "and found it one of the few cities in the Mid-South territory not seeking a defense project of some sort, and singularly free of much desire for one." [15]

In a conversation with former Paris Mayor John Reeves, Reeves told Tucker "The town's businessmen felt generally that there were as many bad features in such a project as there were good features and that the city would not go actively into the market for a project." [16]

The late Charlie McClure, who worked for decades for the Spinks Clay Co., believed it was important to preserve the camp's history. To that end, he collected memorabilia and information about the camp and took numerous photographs of the property over the years, which are still kept at the Spinks headquarters on the former camp's grounds.

According to a history of the camp written by McClure, the initial phase of the camp occupied 1,680 acres, 1.5 miles wide north and south and 2.5 miles east and west. There were a total of 400 buildings, ten miles of asphalt roads, five miles of railroad, a post office, two chapels, a hospital, service club, library, theater, guest housing, hydrogen generator plant and a state-of-the-art water supply system. [17]

Total cost to the government for camp construction was $11,708,604, with the 75-foot-tall flag pole in the center of the camp's

---

[14] Bruce Tucker, "Balloons Boom in Paris, Tennessee," *Memphis Commercial Appeal*, January 18, 1942, section 4.

[15] Ibid.

[16] Ibid.

[17] Camp Tyson history written by Charlie McClure, not dated.

parade field costing $2,754 alone. The original camp provided quarters for 535 officers and 8,356 enlisted men. [18]

Unlike other public works projects of the era, there were no deaths recorded in the construction of the camp, with the only major accident being when a worker lost an arm, *The Paris Post-Intelligencer* reported. [19]

The camp was officially named Camp Tyson, in honor of Brig. Gen. Lawrence David Tyson of Tennessee, who commanded soldiers in the Spanish-American and World War I. He also became owner of *The Knoxville News-Sentinel*.

Even before the camp was completed, things began happening rapidly.

By January 29, 1942, the camp's Barrage Balloon Training School (BBTC) was already taking shape. According to an article in *The Paris Post-Intelligencer* of that date, 52 officers and 246 enlisted men, along with seven officers of the Barrage Balloon Board, had been added to the permanent personnel at Camp Tyson. "This is in addition to the officers and enlisted personnel of the Station Complement who have been at Camp Tyson since early in December." [20]

Purpose of the BBTC School was to train officers and enlisted men in technical matters "pertaining to the handling and flying of the barrage balloons." There would be a general course and two special courses for officers, the former lasting six weeks and the latter four weeks. The student capacity of the school is 240 officers and 340 enlisted men. [21]

The Barrage Balloon Board consisted of seven officers who were overseers of the school.

TOO MUCH LUXURY?

The Army Hospital was in operation at the camp by Feb. 2, 1942. The officers and enlisted men of the Medical Detachment had moved into their permanent quarters at the hospital. The officers' quarters was well-furnished since they were temporarily using furniture intended for use in the nurses' quarters. "Most of the officers report that they

---

[18] Spinks Clay company web site, "About Us: Camp Tyson" (4/2/2009).
[19] Paris P-I, sesquicentennial edition, 4C.
[20] "Camp Tyson BBTC School Takes Shape," *The Paris Post-Intelligencer*, January 29, 1942.
[21] Ibid.

did not sleep very well at first due to the fact that they were not accustomed to luxurious mattresses which are a far cry from the bunks to which they had become accustomed," according to a newspaper account from that time. The luxury was to be short-lived, however, since 49 nurses assigned to the camp were due to arrive within the month. [22]

On February 4, 1942, *The Paris Post-Intelligencer* announced, "Camp Personnel Increased to 2,000 By New Arrivals." A total of 983 men and seven officers from the 302nd Battalion had arrived, boosting the number stationed there to 2,000. "The new arrivals came from Army posts scattered all over the United States and were more or less selected for their new tasks because of their training, ability and leadership," the article said, adding that the full complement of soldiers was to arrive at the camp by March or April. [23]

On February 9, 1942, Colonel Robert Arthur, who was then the commanding officer at the camp, announced he had just returned from Washington, D.C. and that the camp would be immediately begin preparing for its first balloons.

That day, he said, "Construction of balloon beds from which barrage balloons will be flown, was started, and the promise that Parisians would see the long-awaited balloons flying over Henry County, within a very short while, was made." Indeed, "the balloons and their auxiliary equipment are en route to Paris, but there will be considerable preparations necessary before it will be possible to fly them," the newspaper said. [24]

Col. Arthur said special interest was being taken in the construction of the balloon beds "for it is from these that actual training of crews and flying of the balloons will be conducted." The beds were to be laid out in 90-foot squares, and so arranged that the balloons "can be stretched, inflated and prepared for the air, each from its own bed. Each 'bed' will have its own winch, its own crew and special equipment. Fifty of these 'beds' are being constructed now," according to *The Paris Post-Intelligencer*. [25]

---

[22] "Army Hospital In Operation At Camp Tyson," *The Paris Post-Intelligencer*, February 2, 1942.

[23] "Camp Personnel Increased To 2,000 By New Arrivals," *The Paris Post-Intelligencer*, February 4, 1942.

[24] Ibid.

[25] "302nd Battalion Begins Services At Camp Tyson," *The Paris Post-Intelligencer*, February 9, 1942.

The U.S. Army Corps of Engineers of West Memphis, Arkansas, awarded the contract to build the balloon beds to Herschel Smith, a general contractor in Bruceton, Tennessee, for his low bid of $21,000, according to his son, Jonathan Smith, who lives in Jackson.[26]

"He was doing considerable construction work at this time in the Tri-Counties area (Carroll, Henry, Weakley) and during the Camp Tyson work, our family moved from Bruceton to McKenzie in Carroll County," Smith recalled. Herschel Smith supervised the job closely, with assistance from a "reliable foreman, a local man" and local labor, although the gravel and cement came from Nashville and Marquette companies, his son said.[27]

Smith said the original contract was to build 28 beds, to be located on the camp but not visible from the main highway. Security was tight during the construction. "My father describes the beds in the following general manner. Each bed was a circle of about 70-feet diameter; in the center of the circle or bed was a rectangular hole, say 3 ft. by 6 ft. poured with concrete, to which was anchored a bolt or a bolt-type apparatus, to which the large inflated balloon was centered or held."[28]

Located around the bed were 8 or 10 circular holes, 2 ft. by 2 ft. (2 1/2 feet deep) made of concrete, from which there were steel anchor rods, used in the actually maneuvering of the balloon."[29]

After the camp was shut down and much of the property was re-acquired by private landowners, many of the barrage balloon beds were plowed up by farmers. However, several of them still remain on property which formerly was camp land, including that of Joe Lankford and John Steele, both of whom live along Steele Rd., between Routon and Paris.

Both Steele and Lankford said they preserved the balloon beds because they believed it was important to do so.[30] Lankford said, "Most people dug them up with dozers, but I wanted to keep some just the way they were so that people could remember them."[31]

---

[26] Jonathan Smith, letter to Shannon McFarlin, April 30, 1992.
[27] Ibid.
[28] Ibid.
[29] Ibid.
[30] John Steele and Joe Lankford, personal interviews with author.
[31] Joe Lankford, personal interview with author, April 21, 2009.

The first balloon was set aloft on February 13, 1942, by members of the 302nd Battalion. Howard Koenen of Murray, Ky., was one of the men who helped with the maneuver (see Chapter Five). According to *The Paris Post-Intelligencer*, the first balloon was seen by residents of the nearby area as "the gigantic gas bag rose several hundred feet above the ground" around 5 p.m. that day.[32]

Despite it being Friday the 13th, the first ascension was a success, with the balloon being named "Clytee of Battery B, The First Barrage Balloon To Go Up In Tennessee."[33] A real mouthful.

Immediately upon the premiere ascension, plans were made for the camp's first visitors' day, set for the upcoming Sunday.

Brig. Gen. John B. Maynard of Virginia officially assumed command of the camp on February 15, 1942. Until his quarters at the camp were completed, Maynard stayed at the two-story mansion home of Sidney and Elna Mandle, located at 505 Walnut Street in Paris. The Mandles' daughter, Marilyn Mandle Dick, who now lives in Knoxville, said her family and the Maynards were not acquainted before the camp opened. "We were just asked to let he and his wife stay there for a bit. They stayed in the room over our garage and we just enjoyed having them there."[34]

SHINY WHITE BARRACKS

The first visitors' day attracted a whopping 3,000 people to the camp. "So eager was the populace of Henry and adjoining counties to see" the camp "that the rains and extremely bad weather of Sunday afternoon did not dampen the ardor" of residents who toured the facility during the afternoon.

According to *The Paris Post-Intelligencer*, even though announcement that the camp would be open for visitors was not made until the newspaper's Saturday edition, by 1 p.m. "a string of automobiles a half

---

[32] "First Barrage Balloon Is Sent Aloft," *The Paris Post-Intelligencer*, February 14, 1942.

[33] Ibid.

[34] Marilyn Mandle Dick, telephone interview with author, November 15, 2010.

mile long was parked waiting the opening of the gate." A steady stream continued through the afternoon, with lengthy waiting lines. [35]

It was necessary to register each automobile, the number of its occupants and to check against use of cameras or weapons. "The visitors were permitted to drive through all areas of the camp site and were amazed at the extent of the post. The rows of shining white barracks, dining halls, administration buildings, etc., made an impressive picture and the cleanliness and order of the while outfit was impressive in spite of the muddy condition of the new ground." [36]

The visitors' days continued to be popular, as they were held regularly while the camp was in operation. Joel Summers of McKenzie has photographs of his parents and their friends at the camp during a visitors' day. In one, a group which includes Virginia Hudson, Carolyn Thompson Clericuzio, Martha and Margaret Adams, Russellina Hilliard Summers, and Kathleen Edwards Mitchell stand in front of an inflated barrage balloon. In another, Margaret Adams, Russellina Hilliard Summers, R.B. Summers and Martha Adams pose in front of a long line of white barracks.

Ralph Anderson of Paris, who worked for the Spinks Clay Co. for 48 years after the war, remembers watching the camp being built. "Our farm bordered the camp property and I was a little boy, so to me that was pretty impressive, being able to see it being built. As a young fellow, I would go over to the camp and play on things, those whirling things over by the sewer system. We probably weren't supposed to be doing that." [37]

HEMP GROWN AT CAMP

Mike Wimberley of Paris said his grandfather, J.D. Kemp, helped build the 95-foot hangar in which balloons were built and repaired. "My grandfather also told us that the Army used to grow hemp out there. He said the government allotted you so much of it. It's funny that hemp is illegal now, but in those days it was legal. I'm assuming the hemp, which is like rope, was used for the balloons." [38]

---

[35] *The Post-Intelligencer*, Feb. 15, 1942.

[36] Ibid.

[37] Ralph Anderson, personal interview with author, August 6, 2009.

[38] Mike Wimberley, personal interview with author, August 13, 2009.

The late Rufus Hastings was one of the many construction workers who helped build the camp — and helped tear it down after it was closed, too.

"We lived on Macedonia Rd. then, which was a short trip to the camp," his daughter, Ruffalene Webb said. "I remember Daddy used to take us there while it was being built and let us see everything. It seemed like they were continuously building out there."[39]

Webb now lives on Hwy 79 between Paris and McKenzie, just a few miles from where the camp was located, in a house which used to be a camp laundromat. "My Daddy and brother-in-law went out there in 1986 and brought it here on a flat bed truck, all the way down the highway. We had to make a few changes to it and added on it, but I think it's pretty great this used to be one of the camp buildings."[40]

BUILDINGS STILL IN USE

When high altitude bombing began to be the norm, barrage balloons were not used as much as defensive weapons. At that point, the camp became a staging area for troops going overseas and parts of the camp were also used as a prisoner of war camp for German and Italian prisoners.

Several Camp Tyson buildings are still on the grounds and are still used today by the Lhoist North America Spinks Clay Co., which presently owns the property. The majestic hangar is now used to sort clay. General Maynard's quarters are still there, along with a couple of motor pool buildings and several office buildings. The once-modern roadways are still there, although covered in weeds and much harder to travel across.

OVERFLOWING WITH KHAKI

But while Camp Tyson was in operation, it was all hustle and bustle around the camp and around Paris. So much so that a soldier sent a postcard from Camp Tyson to his parents in Syracuse, telling of his experience. On the front of the post card was a photo of several soldiers dwarfed by a huge, inflated barrage balloon, with the heading "Barrage Balloon Training Center-Camp Tyson-Paris, Tenn."

---

[39] Ruffalene Webb, personal interview with author, September 3, 2009.
[40] Ibid.

His message to his parents, which was sent October 23, 1942: "Weather much pleasanter today but still quite cool. Am spending night near this army camp. The town is overflowing with khaki. Love-Whit." [41]

---

[41] Postcard posted on Flickr by Tom Barnes.

AS IF THEY WERE OURS

Chapter 3

"They Just Came Out and Destroyed Our Farms, Our Barns, Our House"

THERE once was a majestic plantation house, one of the oldest homes for miles around. It was called Bowdenville, where four generations of the James Jefferson Bowden family had lived. The massive, two-story colonial home was stately, standing as a symbol of all of Henry County's earliest pioneers.

Today, it is nothing but a memory — a symbol of eminent domain and the negatives that term can represent.

For most, Camp Tyson was positive in every way.

But for others, the construction of Camp Tyson meant heartbreak and loss and lingering feelings of resentment. Numerous families watched their homes bulldozed and set afire, along with the crops which represented their livelihood.

They were given little warning by the U.S. government that their property was being acquired for the camp and had to hustle to move their belongings from their homes.

Having their homes taken was especially hard for them to accept when, only three years later, Camp Tyson was shut down.

When the decision was made to close the camp, the families were given first chance to purchase their properties back. Some of the lucky ones were able to scrape together money to buy back their family farms. Others were not so fortunate.

The first phase of Camp Tyson had already been constructed, but the U.S. Army planned an expansion of the camp to build more balloon beds in 1942. Notices were sent out to families along what is now Steele Rd., Gate 3 Rd. and other nearby roads that their properties would be needed for the expansion.

Families were paid for their property, but descendants today doubt whether the government paid a fair market price. Joe Hill of Union City grew up in Henry and said the 300-acre farm owned by his grandfather, Clyde Looney, was purchased for the camp.

"I remember Daddy saying he was paid $26,000 for his land, which was a pretty cheap price for it," Hill said. "It was real flat land, prime land, and it is where the camp ended up building their hospital." [42]

Despite not receiving adequate compensation for the property, Hill said, he does not recall hearing of any resistance by Looney to the sale. "The first section of the camp was already built by then and he decided that when the soldiers hoisted the balloons in the sky, it scared the cows. So, there really was no resistance from him when they offered to buy him out." [43]

Hill said his grandfather was not afforded the opportunity to purchase the property back after the camp closed. "I didn't hear that he wanted it back by then," Hill said. "He was in his 70s by then and had a house in Henry. His home place had already been destroyed and I'm not sure he would have wanted to move back there." [44]

The Steele family were among the lucky ones who were able to acquire their land back, but even today, their memories of that time period are sour.

"When they tell you that your property is going to be taken and they don't even ask, they just give you a date to be off your own property, well, that's pretty hard." Ruth Steele Robbins said. [45]

"Our crops were still in the field and everyone worked as hard as possible to harvest it," Robbins said, recalling that many concerned people from Paris came out to the Steele farm to help with the harvest by the government-imposed deadline. The government, she said, "just came out and destroyed our farms, our barns, our house. We were paid something for it, but whether it was fair market price, I really don't know." [46]

Her family moved in with her sister's family until her father could build them another home, she said. "There were 10 of us in the same house, so it was very crowded." And to make things more difficult, she said, lumber was hard to come by with the war ongoing and building materials being rationed. [47]

---

[42] Joe Hill telephone interview with author, Dec. 3, 2010.

[43] Ibid.

[44] Ibid.

[45] Ruth Steele Robbins, personal interview with author, May 18, 2009.

[46] Ibid.

[47] Ibid.

"We Were Not Happy About the Move"

Members of the Steele families gathered around the dining room table of their Steele Rd. home one morning in the spring of 2009 to recount how Camp Tyson changed their lives forever. The looks on their faces as they recalled watching their homes being set afire told the whole story.

"We were just devastated," John Steele said. "We knew we had to get out, but we had work to do. People from Paris came out here to help us gather our corn."[48] No, he said, "We were not happy about the move."[49]

Robbins recalled, "They burned all our barns, our houses. They came in with big trucks and burned it all. You could see the fire for miles around. My family had a big, beautiful orchard and it was all bulldozed."[50]

The government, John Steele said, "just wanted it cleaned off. They didn't care how. We worked around the clock, day and night, to get those crops in."[51]

Some families did fight it, but were unsuccessful.

Nancy Cate of Paris recalled that Gus and Inez Barfield owned land that the government wanted for the expansion. "Gus decided to fight it. He wasn't for the idea of eminent domain. He took them to court. Five men were appointed to survey the damage to the property and this went on into 1943. Well, Gus fell dead on the property from the stress and Inez didn't want to follow through."[52]

Another neighbor, Cate recalled, also said he would fight it "and a black car picked him up and took him to Nashville. When he got back, he set his furniture off on the side of the road and went packing."[53]

Lou and Bettye Carter live at the intersection of Steele Rd. and Gate 3 Rd., on land that her family was able to purchase back from the government. The Carters have collected a large amount of memorabilia from the Camp, including maps of the property, pillow cushions, matchbooks, and other souvenirs. On their property is a tree in which

---

48 John Steele, personal interview with author, May 18, 2009.
49 Ibid.
50 Ruth Robbins interview.
51 John Steele interview.
52 Nancy Cate, personal interview with author, May 28, 2009.
53 Ibid.

a lovesick soldier carved his initials and the initials of his girlfriend, surrounded by a heart.

The Carters also have a copy of the U.S. District Court order in which Judge Marion S. Boyd granted the U.S. government the right to acquire the properties amounting to 4,365 acres for the expansion of Camp Tyson. It directs that "possession of said lands be surrendered and delivered to the United States of America on or before the 19th day of October, 1942." [54]

MOWING THE CAMP FOR $1 AN HOUR

Joe Lankford recalls that his grandfather, the late John Steele, was hired by the federal government to mow the camp, even after their land was taken.

"Granddaddy ran four teams of mules and he mowed that entire piece, thousands of acres, that way. He had a contract with the Army. He was paid a dollar an hour," he said. [55]

Lankford has copies of his grandfather's invoices for the work. One invoice, dated July 17, 1943, seeks payment "for the services rendered in the furnishing of teams, driver and mower from May 20 through June 30, 1943." [56]

For 396 hours, at a dollar an hour, Steele was paid $396.

Lankford also has an old photograph of Steele with his team of mules.

"Granddaddy owned 600 acres and he bought it back after the war," Lankford said The Lankfords now live on the family farm on Steele Rd. [57]

His late mother's home is something of a reminder of days gone by. It was constructed from an old army barracks and is adjoined by the only remaining complete balloon bed still remaining in Henry County.

"To build mother's house, they couldn't get any lumber or materials after the war, so they built the house out of the barracks," Lankford

---

[54] Copy of court order, property of Bettye and Lou Carter.
[55] Joe Lankford, personal interview with author, April 21, 2009.
[56] Copy of invoice, property of Joe Lankford.
[57] Joe Lankford interview.

said. "They moved the barracks and used it for the house. The main living area of the house is the barracks and we added on to it later." [58]

In the field right next to the home lies the circular balloon bed, with the 'tie-downs' still anchored into the ground. Like John Steele, Lankford believes that despite the family's experiences with the camp, it was important to preserve as much of the camp as they could.

"A lot of the farmers around here plowed the 'tie-downs' up," Lankford said, to make way for farmland. "...we did with some of them, too, but we kept this balloon bed preserved. It's the only complete one still around." [59]

"THE GOVERNMENT CAN TAKE"

The late Rexie Smith of Routon was one of the workers who helped build the camp — and in the process helped tear down some of the private properties. His daughter, Nelda Pinson, recalls that now with sadness.

"Daddy's first job was to assist in tearing down. I learned that the beautiful Bowden home was the center of activity for the new camp. Our grandmother being a Bowden from Routon and a direct descendant of the family that lived in that home makes me sad to know my Daddy probably helped in destroying it." [60]

Pinson said, "I've often wondered why they couldn't just leave it standing and use it as some sort of headquarters. But I've found that the government can take, no matter how much you yank and pull." [61]

---

[58] Ibid.
[59] Ibid.
[60] Nelda Pinson personal interview, Oct. 9, 2009.
[61] Ibid.

Chapter 4

"They Didn't Feel Like Strangers"

TRAFFIC was bumper to bumper on the roads leading into Paris. The streets of Paris were crowded with men, walking from door to door in the residential neighborhoods, looking for places to stay. A room, a part of a room, a piece of floor — anyplace to set their heads down at the end of the day.

Even before the first public announcements were made, people knew by word of mouth that an army camp would be opening near Paris. The camp had no name and no one yet was aware that it would be the only one of its kind in the United States. All they knew was that rural Tennessee had been hit hard by the Depression and they needed jobs.

An army camp meant work — lots of work and good-paying work — and people from miles and miles around showed up on the doorsteps of the people of Paris, looking for any small place where they could call home, no matter how temporarily.

Overnight, Paris became a boom town.

It actually came in two waves. First, the construction workers arrived, even before the camp started being built. Then after Camp Tyson opened, the soldiers and officers began knocking on doors, looking for places where their families could stay.

U.S. Census figures for the period show the population for Paris at the beginning of the Depression to be 8,164. With times hard, many rural Southerners migrated North during the Depression years, which is reflected in the sharp population decrease in Paris to 6,395 by 1940. With the Camp Tyson boom, however, the population of Paris made up for its Depression era loss and then some, increasing to 8,826 by 1950. [62]

As the camp grew, the government began building "government housing": or what people in Paris still call "the Camp Tyson houses." They are immediately recognizable: identical, compact, brick structures which still provide good homes for people on Jackson Street, East Washington Street, Lake Street, Chickasaw Rd., and several other

---

[62] U.S. Census Figures, Paris-Henry County Chamber of Commerce.

neighborhoods. And those housing projects provided even more construction jobs.

But before the extra housing was provided, places for the workers and the soldiers' families to stay were at a premium.

The people of Paris responded to the need, opening their doors to these strangers, offering every nook and cranny of their homes. The physical closeness of the quarters led to closeness of a different sort — lasting friendships developed from the Camp Tyson experience, friendships that endured for years long after the war was over.

Local households charged rent, which in most cases was a minimal amount. But the rent, while welcome and needed in tough economic times, could not fully compensate for the lack of space and privacy that the crowded accommodations must have meant for local families.

Creative ways were found to house the new guests. Back porches, rooms or portions of rooms, attics, sheds, and, in at least one case, a tree were fashioned into housing. Former Camp Tyson soldier Howard Koenen of Murray, Kentucky, recalls that when he and his wife were newly married, they rented a place in Paris at which the homeowner also was renting the tree in the backyard to soldiers. "The government had a law that you could only rent so many rooms of your house and only charge so much," Koenen said, but the homeowner got around that by also renting his tree. [63]

Today, the people of Paris say they did not even notice how overcrowded their homes were during that period. It was World War II, after all, and sacrifice was expected and promoted. Many housewives actually enjoyed sharing their homes with the soldier's wives — and in some cases, their children, too. It provided much-needed company during anxious times.

THE MAYOR MAKES AN APPEAL

Two people who had a bird's-eye view of the housing situation and the steps that were taken to accommodate the influx were Jack Lasater and Martha Lasater Andrus, whose father was the mayor of Paris during the period.

Mayor John P. Lasater, Jr. and his wife, Lauriene, were a team, working together to see to it that the soldiers and their families had places to stay.

---

[63] Howard Koenen, personal interview with author, October 30, 2009.

AS IF THEY WERE OURS

City officials were unhappy that many officers were already finding housing in other towns. A February 4, 1942, article in a local newspaper lamented that 40 officers already had settled in Mayfield, Kentucky, at least an hour's drive from the camp. [64]

Mayor Lasater placed an appeal in the newspaper, asking the people of Paris "to come to the aid of the incoming army officers...He asks that any person who can provide living quarters for anyone" to call city hall (phone number 121). [65]

And in the days that followed, a local editor added his two cents, noting that now that the government had designated Paris as a "defense area," providing housing should be an easier proposition since federal loans would be available for new construction. "The considerable influx of permanently located army officers," the editor wrote, "must have houses and apartments for their families...The question of whether Paris is ready for an expansion is now up to our own citizens. We have been suffering from a shortage of houses for years, even before the war emergency." [66]

The editor noted that Paris was especially hard hit from the depressed real estate market and the time was now for action. If Paris should fail to take advantage of the 'defense area' designation, he said, "We may as well notify Camp Tyson officers to find living quarters in Mayfield, Murray, McKenzie or somewhere else..." [67]

But the editor said he believed the people of Paris would do the right thing. "Since Paris has been officially designated as a defense area, our opinion is that building activities will get under way by the time the weather breaks." [68]

Jack Lasater, who now lives in Siloam Springs, Arkansas, said the housing shortage was something that both his parents took to heart, opening the doors to their own home at 314 Jackson St.

"Dad realized there was not going to be adequate housing for the married couples early on and I remember we had at least three couples

---

[64] "Forty BBTC Officers Have Already Settled In Mayfield," *The Paris Post-Intelligencer*, February 4, 1942.

[65] *The Paris Post-Intelligencer*, February 3, 1942.

[66] "Paris Defense Area," *The Paris Post-Intelligencer* editorial, February 5, 1942.

[67] Ibid.

[68] Ibid.

staying with us at various times," Lasater said. "Both Dad and Mother would issue appeals to their friends to let people stay with them."[69]

His sister, Martha Lasater Andrus, who now lives in Boise, Idaho, remembers that their mother would go all out in the quest to make the soldiers and their families welcome.

"I remember every Sunday if there were soldiers at our church, the First Presbyterian, that mother would invite them home with us for Sunday dinner," Andrus said. "Well, one Sunday, no one came and on the drive home, Mother drove around the court square and she saw two soldiers standing on the post office steps. She stopped and invited them to come with us for dinner. Well, you could tell they didn't really want to get in a car with these people they didn't even know, but they did. And after everyone ate, Mother took them back to the camp."[70]

Andrus also recalled her mother made sure a newly-wed couple had at least some sort of celebration in war times. "There was a soldier stationed at the camp and he sent for his sweetheart to come and get married. She came from New York and he met her at the train station and then they went to the Justice of the Peace to get married. Mother couldn't stand it when she heard they were at city hall getting married, so she baked them a cake and had some punch for them at our house. It was these strangers, plus Mother, Jack, who was just a little boy then, and I at this little reception, but Mother couldn't stand not having a little celebration for them."[71]

Andrus said her father "just thought it was the right thing to do, to have the families living in peoples' homes, but during that time, everyone was involved in the war effort. Everyone rolled bandages, everyone gave blood. I remember when I was a senior in high school, a few of us even worked for the Office of Price Administration, helping the grocery stores set their prices. These were things you just did."[72]

"JUST BUMPER TO BUMPER"

Virgil Wall remembers how quickly things changed in Paris. He was visiting his uncle in St. Louis, in the summer of 1941, when his mother wrote him a postcard saying that an army camp with barrage

---

[69] Jack Lasater, phone interview with author, February 15, 2010.
[70] Martha Lasater, Andrus phone interview with author, February 28, 2010.
[71] Ibid.
[72] Ibid.

balloons was going to open. "Of course, no one knew what that was," he said.[73]

"By the time I came home two weeks later, a lot was going on," Wall said. "There was a tremendous influx of people coming into town to help with the construction and there was nowhere to put them. The city was appealing to everyone to help."[74]

Wall said his family divided their home, "making it into a duplex and we moved to one side of the house and rented the other side out to families. A whole family stayed with us, a man, wife and son, and later we had individuals staying. We rearranged the top floor later."[75]

His father owned other houses, which he rented out, and constructed a small, two-room house to rent.

One family moved in with the Walls, more or less permanently. "They stayed with us for a year or more and became family friends. We continued corresponding after the war." Wall recalled that when the "Camp Tyson houses" were built by the federal government along Chickasaw Rd. that section of town "wasn't populated at all out that way. Actually, that was supposed to be the entrance into town from that direction."[76]

The wife of one of the soldiers that lived with his family "used to drive her husband to the camp and I liked to ride out with them. I remember it was just bumper to bumper with cars all the way out there. And the road there was not very good. It was muddy and dusty, depending on the time of year. Cars used to get stuck in the mud all the time on the way out there."[77]

His wife, Jo (Randle) Wall, said her family also rented rooms. "Our house was close to town, where city hall is now, and we had several soldiers stay with us. We rented out our bedrooms."[78]

Paris was "bustling" during that time period, she said. "I worked at Fry Drugs on the court square when I was 16 and we were so busy that we stayed open every night until 10 or 11 p.m."[79]

---

[73] Virgil Wall, personal interview with author, May 12, 2009.
[74] Ibid.
[75] Ibid.
[76] Ibid.
[77] Ibid.
[78] Jo Wall, personal interview with author, May 12, 2009.
[79] Ibid.

Bryant Williams, whose family has owned *The Paris Post-Intelligencer* for generations, said with Camp Tyson opening, "Paris really grew, seemingly overnight."

His family housed a young couple, "a 2nd Lt. and his wife. We had a little house on E. Wood St. and it was me, my wife and our son, Bill, who was seven then. We moved into the basement and we let the young couple live upstairs. And it was only a three-room house!"[80]

A NEIGHBORHOOD FULL OF SOLDIERS

As a little boy, Franklyn Thompson lived with his parents and sister Jane in a home at 403 Edgewood St., Paris. "I remember we had an unfinished attic and sealed it up and rented it to a Lieutenant and his wife. I remember the soldier brought us a whole pie as a gift one time and that was really appreciated by us kids."[81]

The neighborhood where the Thompsons lived was crowded with soldiers and their families. For several blocks in that area of Paris, seeing soldiers was a common — and rather comforting — sight.

One block over, at 713 Park Street, the home where Jeannette Snow McFarlin of Paris grew up was crowded to the rafters. "I remember at one time we had five families living here," she said. "That was during the construction phase and we just divided up the house to give each family some privacy. I stayed out on the little room on the back porch, but when we rented to a soldier and his wife, I had to give that up. That's where they stayed."[82]

All the families ate meals at different times, she said, with each housewife having access to the kitchen. "I guess it was crowded, but we really didn't think about it." McFarlin said, "We corresponded with the soldier and his wife for several years after the camp closed."[83]

Across the driveway at 711 Park Street, where the Burtons lived, the story was the same. "We had a three-bedroom house, so we did have some extra room," according to Mary Alice Burton Campbell of Paris. "When the construction workers came, we gave up two rooms

---

[80] Bryant Williams, personal interview with author, April 7, 2006.
[81] Franklyn Thompson, personal interview with author, February 12, 2004.
[82] Jeannette Snow McFarlin, personal interview with author, May 25, 2009.
[83] Ibid.

to them. My sister and I moved into the living room. My brother slept in the breakfast room."[84]

Later, soldiers and their wives stayed with them, too. "They became part of the family and we kept in touch with them until mother died in 1989. Everyone opened their homes. It was just part of the war effort. No one was scared to have strangers in their homes. They didn't feel like strangers, especially when you think about the part they were playing in the safety of the country. Times were different then."[85]

And one more house over, at 709 Park St., where the Rileys lived, even more soldiers and their families stayed.

"It was a lot of fun when the soldiers would come home in the evenings," McFarlin said. "I remember all us kids, the Burtons, Rileys and me, would go out into our driveways and salute them when they got out of their cars."

After school, the neighborhood kids would gather in the front yards, watching the barrage balloons flying at Camp Tyson some 10 miles away. "The trees weren't as high then in our neighborhood, so you could see the balloons clearly," McFarlin said. "All us kids would count how many we saw and each kid always seemed to have a different number for some reason."[86]

A few blocks away, at 325 Head Street, the Hedges family opened their home, too. "We had four construction workers in one bedroom and in our living room," Bobbie Hedges Parker of Paris said. [87]

Parker is still in touch with the soldiers' families which came later. "There was Chuck and Louise Day and when mother died they came back to our house to help Daddy. Another family was Bill and Juanita Wilsman. He was from Pittsburgh and she was from Arkansas. They used to come through Paris all the time after the camp closed. They were just family to us."[88]

The neighborhood was so crowded with soldiers' families, she said, "The traffic was just bumper to bumper. You'd have to have help getting down the street. It was just one car after another."[89]

---

84 Mary Alice Burton Campbell, personal interview with author, January 13, 2010.
85 Ibid.
86 Jeannette Snow McFarlin interview.
87 Bobbie Hedges Parker, personal interview with author, October 15, 2009.
88 Ibid.
89 Ibid.

Julia Hutson remembered that well. Her family lived in the same neighborhood, at the corner of Edgewood and Park. "This town was swarming with activity," she said. "Paris was not prepared for the sudden expansion, all at once."[90]

A Future Astronaut

Among the soldiers who moved to the area because of Camp Tyson was Leroy Gordon Cooper, Sr., father of the future Mercury 7 astronaut Leroy Gordon Cooper, Jr., who became affectionately known to the American public as "Gordo."

Cooper Sr. settled his family in Murray, Kentucky, a town 22 miles north of Paris. Cooper Jr. graduated from Murray High School and his former classmates recall that he seemed "reluctant to cultivate close friendships" while at Murray, which they surmised was because "he was an 'army brat' accustomed to moving from place to place and school to school.'" His former classmates did recall that he was active in the Murray Boy Scout troupe and diligently pursued merit badges.[91]

Knocking on Doors

The Hutson's door began knocking "with men asking us if they could stay here. And, of course, I couldn't say no. I moved things around the house and our family lived in three rooms, and we had three men staying in the front room, three men in the back room."[92]

After the camp was built, the Hutson home was opened to the soldiers and their families, too. "We had rooms full of people living with us. And it wasn't really a strain — everyone was one big, happy family. Those really were good memories." Julia and her husband, Solon, had two young daughters of their own by then, "but no one was afraid. It was actually good company for me, Joyce and Patsy during the day."[93]

Hutson was kept busy making sandwiches for the workers, which evolved into a side business making sack lunches for the troop trains that came through Paris. (See Chapter 8).

---

[90] Julia Hutson, personal interview with author, September 16, 2004.
[91] "Murray Classmates Remember: Astronaut Gordon Cooper," *Montage*, Murray, Kentucky, June, 2010, page 7.
[92] Ibid.
[93] Ibid.

But beyond supplying sack lunches for the troop trains, Hutson supplied meals for soldiers and their families from her home. "We had this big house and men were stopping here asking for rooms. We rented the attic, the back and front rooms and then they wanted to know if I'd make their lunches. Then three men who were staying up the street asked, so I was making lunches for nine men altogether." [94]

Hutson said her husband made a large table that could seat ten people so that everyone could eat their meals together. "It was like a boarding house. And it was fun." Hutson said she couldn't recall how much her family charged to rent the rooms out or for making the meals. "It wasn't much and sometimes I didn't charge anything at all," she said. [95]

"It was a different atmosphere then," Hutson said. "People were so desperate. I remember a couple came here just after dark. There was going to be a big shipment overseas the next day. The soldier would be leaving and he had a wife and baby who needed a place and he heard we might have a room, so they stayed here. That time, we had two couples here already and we didn't really have a spot for them, but I couldn't turn them away and I didn't charge him, either." [96]

Hutson said her family had people staying in their home from Oklahoma, York, Pennsylvania; New York, Illinois and other places she couldn't recall off-hand. "It was real interesting to hear about their lives. The young wives would be here all day and they were just like my sisters." A life-long gardener, Hutson's large vegetable garden was a big help during that time — and a source of fascination for her boarders. Hutson taught several of the soldier's wives about gardening while they were there "and I remember one girl from New York who had never tasted corn out of a garden. She said she never knew it was that good." [97]

In the evenings and weekends, when the soldiers were there, she said, "We'd all sit out on the porch and talk." When the second phase of construction at the camp was underway, she said, the neighborhood once again became busy with workers seeking a place to stay. "I remember Solon and I were out one night and we came back around 8

---

[94] Ibid.
[95] Ibid.
[96] Ibid.
[97] Ibid.

o'clock and one of the workers who had stayed here before was sitting on the porch, waiting for us. He wanted to stay with us again." [98]

She said she regrets never taking any photographs of her boarders, but said "we used to get letters from the people who stayed here, for a long, long, time after they left." The families stayed in touch over the years, she said. "We'd get letters from the wives and they'd send us pictures of their children. One of the soldiers sent me a letter from Italy, where he was sent after Camp Tyson. And another soldier sent me a handkerchief. Those were special times." [99]

"WE'RE GOING TO LIVE WITH YOU"

Late one evening the Morris family on Walnut Street heard the doorbell ring. "Mother opened the door and there was an officer standing there with his wife. He said, 'We're going to spend the night with you tonight and then we're going to live with you," Carol Morris Bowden recalled. [100]

The officer was very persistent, Bowden said. "He just said, 'We're going to do this.' He said he and his wife had been looking for a place to stay all day long and they were exhausted. His wife was recovering from an operation and they'd been everywhere looking." [101]

Bowden, who now lives in Montgomery, Alabama, said their home was small "but Mother had a heart of gold. She said they could stay that night, but that we couldn't keep them indefinitely." That night, Bowden slept on the sofa and the couple took her room. "They were at our house for a year," she laughed. "They became a part of the family." [102]

Colonel Moye, she said, "was the crustiest person you've ever seen, but he was a good man and we all became good friends. His wife's health was not good and she was on crutches there at first and we still have photos of them. Even after the war, they still visited us and we went to visit them in Atlanta." [103]

---

[98] Ibid.
[99] Ibid.
[100] Carol Morris Bowden, telephone interview with author, March 10, 2010.
[101] Ibid.
[102] Ibid.
[103] Ibid.

LASTING MEMENTOES

Red Boden said his family also housed soldiers in their home on Blanton Street in Paris. "We became especially close to Wilbur Carpenter. He was the last officer at the camp and helped close the place. He and his wife came back to visit us years later. Wilbur was able to retrieve some paintings from the officer's club which he gave to my mother and father." Boden still has the paintings, one of which depicts a deflated barrage balloon. "That one is especially nice for me because you don't often see the balloons deflated like that. There are a lot of pictures of the balloons up in the sky, but not many like this," he said. [104]

COMMUNICATION PROBLEMS

At least two local families had experience with Cajun families from Louisiana, which made for some early communication problems.

"We had a soldier and his wife from Louisiana. I remember the wife's name was Dorothy, but I can't remember their last name right now," the late Bettye Craig Hill recalled. [105]

The Craigs lived near the camp, in the little town of Henry, and it wasn't uncommon for soldiers to appear in their yard, she said. "You'd be anywhere in your yard and they'd just appear. You'd be gardening and they'd appear behind a bush. They didn't frighten us, though, because they were soldiers." [106]

Her father, Hurley Craig, was a trackman for the L&N Railroad and worked along the track from Paris to McKenzie, which left her mother, Peolia, at home alone with the children most days. "At first, Mother said no way would we rent out a room to the soldiers, but then she changed her mind. It actually was a lot of company." [107]

As African-Americans, the Craigs were fascinated with their Cajun guests. "They had an accent different from ours. Their behavior was different from ours. They had a lighter color than us and their voices sounded different from ours. But we could communicate." [108]

---

[104] Red Boden, personal interview, July 21, 2009.
[105] Bettye Craig Hill personal interview, March 30, 2010.
[106] Ibid.
[107] Ibid.
[108] Ibid.

A couple named Adam and Eve Thibodaux lived with the family of Rebecca Hill Goins, right outside the camp. The Hill's farm was acquired by Camp Tyson for the second phase of construction, but in the early years of the camp's existence, the family housed soldiers and their families.

"We thought it was funny that we had an Adam and Eve living with us. They were Cajun and spoke with a real accent," Goins said. "Eve stayed with us most of the time while her husband was at the camp and she mostly spoke French. I have to say it wasn't that easy to communicate sometimes. We had to use sign language. I remember they'd never seen a persimmon tree before." [109]

Construction workers lived with them, as did a soldier and his wife named Carl and Colleen Holcomb of Brazelton, Georgia.

"We stayed in touch with all of them. I still talk to Eve on the phone and we still have trouble understanding each other," Goins said. "And Carl Holcomb became a preacher and he preached at my father's funeral years later." [110]

Eve Thibodaux, who lives in Choctaw, Louisiana, still has a thick Cajun accent and remembers the Hill family well.

"My husband could speak English better than me. He learned in the Army," she said. "I stayed with the family. I never went anywhere. Adam could walk to camp. We didn't have a car. They were nice people. O Lord, they were so good to me." [111]

Carl Holcomb, Jr. and his wife, Colleen, stayed at Camp Tyson for two years and lived with three different families while there. "I was with the medical corps and I helped open the hospital, with Sgt. James Wilson, who married a local gal and stayed in Paris." [112]

His wife had their first child while they were in Paris and worked at a laundry business in town. "That was the only thing she could get. Everything else was full," he said. Even with a nomadic existence of moving from place to place in Paris, he said, "We were so young, we didn't care where we lived. We didn't care if we lived in one room; we enjoyed it."

Holcomb said the experience was worthwhile for the friendships that were made. "We visited in Paris almost every year for a long time,

[109] Rebecca Hill Goins, personal interview with author, October 2, 2009.
[110] Ibid.
[111] Eve Thibodaux, telephone interview with author, September 16, 2009.
[112] Carl Holcomb, Jr. to Susan Gordon for Tennessee Historical Society 'Home Front' project, April 17, 1992.

and since I became a minister, I've come back twice to perform funeral services, once for Mr. Hill and another for Mrs. Barfield — we stayed in her home, too." [113]

Working at the camp could benefit the families he stayed with during war times, too, he said. "I remember I was on permanent KP at one point. I was telling the cook that the family I was staying with couldn't get sugar ration stamps and he gave me a box of sugar and said not to say anything about it. But I felt good that I could help out the family because times were pretty tough for everybody then." [114]

BISCUITS TO DIE FOR

The family farm for Mary June Smallwood Sinnema was very remote, especially during the 1940s, way out on Guthrie Rd. But even though it was miles from the camp, the Smallwoods hosted construction workers. "We had one room with a two double beds where they would stay," she said. "They just loved my mother's breakfast. She'd make them homemade sausage, biscuits to die for." The workers told her, "We'll pay you to make our lunch, too, so she'd make them lunch and wrap it in wax paper for them." [115]

She also can remember watching the sky for the balloons when she was a child. "We were pretty far from the camp, but we could see the balloons. The sky was peppered with them. Sometimes it would seem the whole sky was full of balloons. And when the sun reflected on them, it was quite a sight." [116]

HOUSES FULL OF SOLDIERS

"The town was loaded with soldiers and their wives," Eddie Moody of Paris recalled. "I remember one gal was pregnant and her husband, a soldier, had been sent off. We took her in at our house. Forty-five years later, she came back to Paris and found my office and said she wanted to pay us for what we did for her." [117]

---

[113] Ibid.

[114] Ibid.

[115] Mary June Smallwood Sinnema, personal interview with author, January 19, 2010.

[116] Ibid.

[117] Eddie Moody, personal interview with author, May 5, 2009.

His parents, Fred and Eula Clymer Moody, already had a full house at their home at the corner of College and Blythe Streets in Paris, but they had several couples staying with them, Moody said.

"We had a little five-room house and we had four couples staying in the garage. They paid what they could pay. People just had their houses full of soldiers." Conditions weren't always the best, he said, but housing was so limited that people took what they could get. "I remember at our house, each couple had their own food and they each had their own slop jar, too."[118]

"Other Cities Didn't Absorb Them Like We Did"

Jeanne Anderson Townsend of Paris, who worked as a secretary for officers at the camp, said other Southern towns were not as welcoming to Army camps as Paris was. "Some people didn't like them because they were Yankees and even some people in Paris were like that. But other cities didn't absorb them like we did. It made for a very comfortable existence for everyone here during the war."[119]

"Whoopie!"

Ruffalene Webb grew up in a former stagecoach station on Macedonia Rd., which made for an interesting experience for her family's boarders. "We were way out in the country. Way out on a gravel road. We had an outhouse and we raised cows and pigs. I think it was pretty different for them."[120]

Her family rented to several construction workers during the construction phase. "They always stayed in the little attic upstairs. There's actually things carved in the wood up in the attic. I remember one weekend we went home and nobody was there. Dad walked up to the attic and came down and said, 'Nobody is here. They're all gone.' Mom let out a big shout and said, 'Whoopie!'"[121]

---

[118] Ibid.
[119] Jeanne Anderson Townsend, personal interview with author, June 2, 2009.
[120] Ruffalene Webb, personal interview with author, September 3, 2009.
[121] Ibid.

As If They Were Ours

Chapter 5

Flying Elephants In the Sky

BARRAGE balloons would fill the sky when the Camp Tyson soldiers performed their daily training rituals. They were so huge and flew so high they could be seen for miles around. They were called "flying elephants," "melancholy elephants," "gas bags" and — because they were filled with hydrogen in combat and sometimes even during training — they were often called "flying bombs."

Henry Countians were fascinated by them. School boys and girls would gaze at them from their school windows and adults would stop whatever they were doing and watch them in awe.

"Back in the 1940s, I would set my farm tractor toward the west and view about ten miles away, oh, maybe 50, more or less, of those balloons flying," the late Frank Brown recalled. "They appeared to be flying, but were tethered to the ground. It was most interesting for this ole country boy." [122]

Barrage balloons were used extensively during World War II as a defensive weapon and are credited with saving the day during the Battle of Britain. Yet to most people, the term 'barrage balloon' is not commonly known. Even Henry Countians, who saw them every day for the years Camp Tyson was in operation, often admitted they didn't really understand what purpose the balloons served.

The history of the "flying elephants" began during World War I when they were used by England, France, Germany and Italy. Their use by the U.S. did not begin until World War II when they were developed by the Army Air Corp and placed in service by the Coast Artillery and Marine Corps. [123]

Their purpose was two-fold: to deter invasion by low-flying aircraft and to be used as a weapon when enemy aircraft flew through the balloons' invisible cables.

As a deterrent, the balloons — which were filled with helium for training purposes — were attached to steel cables that were raised or

---

[122] Frank Brown, email communication with author, April 17, 2006.

[123] James R. Shock, *The U.S. Barrage Balloon Program* (Bennington, Vermont: Merriam Press, 1995), page 5.

lowered by motorized winches. The balloons could be used to protect Allied strongholds by forcing enemy planes to fly at higher altitudes; thus any element of surprise would be lost.

As a weapon, the barrage balloons could be used in this manner: enemy planes would be unable to detect the invisible cables holding them aloft and would fly through them, catching the balloons in the process. The balloons — filled with hydrogen — would then fly into the engine of the enemy plane, causing a deadly explosion.

Once the U.S. decided to go ahead with its own barrage balloon program — the planning of which began even before the Pearl Harbor attack — the key decision was where the balloon training center would be located.

Planners determined the site could not interfere with commercial air traffic or military and naval air operations. [124] Camp Davis in North Carolina was considered — and a temporary operation was established there — but its permanent use was nixed because it did not meet all the criteria. Several soldiers who eventually were assigned to Camp Tyson began their service at Camp Davis, including James Wilson and Cal Orris of Paris and Howard Koenen of Murray, Kentucky.

The choice was then narrowed down to two sites: Danville, Kentucky, and Routon, Tennessee, near Paris. "Danville was the better location from all standpoints except the cost, which was $300 to $500 per acre. The Paris site could be purchased for $50 per acre and consequently this location was purchased..." [125]

The Evening Herald reported in December of 1941 that the U.S. Army had decided on Paris for the camp. "To avoid being caught napping, army officials have set aside a one-thousand acre tract of land seven miles southeast of Paris, Tenn., where officers and enlisted men can be thoroughly trained and formed into barrage balloon units. Finishing touches are being put on a school which will have six and twelve-week courses in the technique." [126]

Initial planning for the Barrage Balloon Training Center (BBTC) included a school, a barrage balloon board, and one battalion. A total of five battalions were to be organized by November 1, 1941 and each

---

[124] Shock, page 22.

[125] Shock, page 22.

[126] Milton Brunner, "Barrage Bags Take To Air; To Guard Not Only Coastal Cities but Many Other Points," *The Evening Herald*, December 3, 1941.

AS IF THEY WERE OURS

battalion was to consist of three lettered batteries and one battalion headquarters battery. [127]

From the U.S. Air Corps, Lt. Col. Robert E. Turley was dispatched to England in June 1941, to study techniques employed by the British and the American military subsequently engaged in reviewing costs and examining materials.

According to a history of the barrage balloon training center published in the Camp Tyson yearbook in 1941, "Barrage balloons were used during World War I with varying degrees of success. Since that war, the United States Air Corps has been experimenting with barrage balloons for many years and had put certain types into construction, but it was not until Great Britain had proved the value of this new defense in the present world war that the War Department ordered the formation of Barrage Balloon units in the United States Army." [128]

The Japanese attack on Pearl Harbor on December 7, 1941, kicked the plans into high gear and "made the barrage balloon program top priority. With the Pacific Fleet severely damaged and Japanese plans unknown, the West Coast was subject to attack at any time. Barrage balloon battalions were immediately dispatched to major defensive sites." [129]

SITE IS CHOSEN

The attack on Pearl Harbor not only accelerated plans for the balloon camp outside of Paris, but increased the number of battalions to be housed there from five to nine. [130]

"On Monday, December 8, 1941, immediately upon the declaration of war against Japan, means were taken to furnish barrage balloon protection for certain vital areas," according to the Camp Tyson yearbook. "The 301st Coast Artillery Barrage Balloon Battalion...was selected by the War Department to be the first unit to move from Camp Davis to an active defense area." [131]

---

[127] Shock, page 23.

[128] *Historical and Pictorial Review, Barrage Balloon School Station Complement, 1942* (Baton Rouge, Louisiana: The Army and Navy Publishing Co., 1942), page 6.

[129] Shock, page 27.

[130] Ibid, page 57.

[131] Historical and Pictorial Review, page 13.

THE FIRST BALLOON IS ALOFT

After months of training, the first balloon to go aloft at Camp Tyson was on February 13, 1942, by the 302nd Barrage Balloon Battalion, Battery B.

Howard Koenen of Murray, Kentucky, remembers that day well. Originally from Connecticut, Koenen was assigned to Camp Wallace outside of Houston, Texas, when Pearl Harbor was attacked. He was immediately dispatched to North Carolina, and then on to Camp Tyson, where he and the other members of his outfit began barrage balloon training.

"We were the ones that put the first balloon airborne," Koenen said. "It was quite an accomplishment and we had a big celebration that day."[132]

Soldiers were trained in how to use two types of balloons — low and high altitude. The very low altitude (VLA) balloons were the ones that were eventually used by the 320th battalion, the only all-black unit to participate in the Normandy invasion on D-Day on June 6, 1944.

Army Tech Sergeant Harold Bostic of Spencer, Indiana, was a member of the 302nd and recalled, "The principle was that if an airplane hit the cable, up under the balloon there was a canister with a long inertia link with a weak spot in it and one at the bottom just above the wench operator, when a plane would hit them, the cable would fire a .45 caliber in those weak spots and separate the balloon from the cable. Up above on the canister there was a seven-inch parachute called a stabilizer that came down. Below was a parachute 54 inches in diameter that was the drag. As the plane went forward, it brought this bomb down, and it exploded and that took care of the plane."[133]

With intensive work by the commanders and soldiers, Camp Tyson took shape. "Now firmly established in its own permanent station, the Barrage Balloon Training Center...and the separate battalions, is rapidly taking its place as an outstanding part of the United States Army," the Camp Tyson yearbook reported. "The rapid progress being made by the personnel was well-demonstrated on Monday, April 6, 1942, during the observance of Army Day. On this date, the camp was

---

[132] Howard Keenan, personal interview with author, October 30, 2009.

[133] Michael Stanley, "Barrage Balloon Training Tech Sergeant Looking Forward To April Honor Flight," *Spencer Evening World,* online newspaper, March 2010.

AS IF THEY WERE OURS

opened to the public and the troops displayed their ability to handle the large balloons and their many other duties to an enthusiastic and appreciative audience. The spectators went away feeling confident that this new branch of the United States Army was prepared to do its share in defending the United States." [134]

The foundation for "all barrage balloon organizations," according to the yearbook, "is the Barrage Balloon School, where both officers and enlisted men are instructed in the mechanics, the tactics and the uses of the barrage balloon." [135]

Since its establishment in the summer of 1941, the school had been "the fountain of knowledge" for both officers and enlisted men of all the balloon organizations, "giving them a wide scope of instruction in this newest of defense weapons. Officers have been given both theoretical and practical courses in gas, aerostatics, operations, communications, rigging and fabric repair, tactics, weather, winches, engineering, etc., while enlisted men have received theoretical and practical courses in gas, rigging, maneuvering, inspection of winch operation, teletype, weather, communications, clerical courses, etc. From this school come the officers and men who float 'the flying elephants.'" [136]

In addition to containing photographss of the officers and soldiers, the yearbook is illustrated with photographs of the training that was employed at Camp Tyson, from classroom work, including laboratory and drafting studies, to instruction in rope tying, winch maintenance, winch operation and rigging.

The yearbook also contains photographs of the soldiers displaying the barrage balloons for the public during the Army Day observance, marching on the parade grounds, installing tents on the grounds, and maneuvering the balloons. It also displays interior and exterior shots of the Service Club, library, cafeteria, the helium purification plant, and photos of the soldiers relaxing over ping pong, reading newspapers and other pursuits.

ANY WIND AT ALL COULD THROW YOU ALL AROUND

A great hazard for the soldiers operating the balloons "was the use of hydrogen for inflation and the hazards should a cable controlling the balloon break, which was a more or less common occasion in the

---

[134] *Historical and Pictorial Review*, page 13.
[135] Ibid.
[136] Ibid, page 17.

early part of the war," according to James R. Shock, author of *The U.S. Barrage Balloon Program*. "A falling cable could create havoc for the handling crew — particularly the winch operator." [137]

Training could be a hazardous operation, according to the late Cal Orris of Paris. Orris was a First Lieutenant with the 861st Anti-Aircraft Artillery (AAA) Battalion and explained that although Camp Tyson did have its own hydrogen manufacturing plant on the base, "We used helium here for safety's sake (while training). You could carry 500 pounds in those balloons." [138]

Orris explained, "We had 12 on a crew and we were dealing with objects that were 12-14 feet in diameter and 50 feet long. Any wind at all could throw you all around. We had a man on every guy rope" to keep them steady. "Once the balloons were in the air, you were all right," Orris said. "The funny part is that no one was ever killed out there. We did have some get away. Gases would expand and the fabric couldn't hold it and it would blow up." [139]

Dudley Atkins of Alabama was a balloon rigger at Camp Tyson. He initially began his service at Fort McClellan, Alabama, and began his basic training at Camp Tyson in Nov. 1942. "They decided they would make combat units out of some of the balloon units," he said.

At Camp Tyson, part of his training was to participate in maneuvers over the Cumberland River near Nashville, and on maneuvers in Louisiana. On the Cumberland, his outfit protected the bridge over the river. On maneuvers, he said, "We would use the VLAs (very low altitude balloons). Four men crews, using hand-powered winches. They were large balloons — it would take 21 men and we had gas engines to put them up and down."

In field maneuvers, he said, "We would use hydrogen gas, but at camp we used helium. You couldn't take a chance with hydrogen. It might start a fire at the camp." During the Louisiana maneuvers, he said, lightning did strike one of the balloons.

Balloons did get away when cables snapped during training, he said, including one that was later found 10 miles from camp. "We could make it fly again after it was lost" by sewing it up and using cement to patch it up, he said.

---

[137] "The U.S. Barrage Balloon Program," page 43.
[138] Cal Ores to Susan Gordon for the Tennessee Historical Society "Home Front" project in Paris, Tennessee, April 17, 1992.
[139] Ibid.

AS IF THEY WERE OURS

After his Camp Tyson training, Atkins was shipped out to Sicily, where "we used the balloons on the beachhead landings. We tried to keep the low level planes from strafing the troops. We lost a lot of balloons out in the seas. The cables broke and the crews got sea sick. We fared so poorly that before long, they broke the outfit up." [140]

As Atkins recalled, field training often took the balloon operators off the base. One of the key training sites was the Scott Fitzhugh Bridge at Paris Landing State Park [141] (rebuilt as the present Ned McWherter Bridge), which is ten miles from Paris.

LAUNCHING BALLOONS FROM BARGES

The late Bill Perkins of Paris remembered training with the balloons at Paris Landing. "We did boat training on the Tennessee River," Perkins said. "We'd launch the balloons off of barges on the river. They wanted you to be able to fly the balloons in all kinds of weather." [142]

Perkins said the balloons — which he said were 40 feet long and 30 feet high — were filled with helium inside of the 90 foot tall hangar at the camp. "You would fill them in the hangar and that way you could patch them in there if they needed repair. Then you'd take them out in the fields. We could fly them a maximum of 5,000 feet, although we didn't usually fly them that high." [143]

Balloons often broke and got loose, especially in high winds, he said. "I can remember finding them out by Humboldt, Van Dyke, Puryear, Whitlock," he said. [144]

Several units which were initially organized at Camp Davis, North Carolina, eventually were stationed at Camp Tyson, including the 302nd (of which Howard Keenan was a member), the 303rd, 304th, and 305th Coast Artillery Barrage Balloon Battalions. The units which were activated at Camp Tyson included the 306th, 307th, 308th, 309th, 310th, 311th, 312th, 315th, 316th, 317th, 318th, 319th, and the all-black 320th Coast Guard Artillery Battalions. [145]

---

[140] Dudley Atkins to Susan Gordon, April 17, 1992.
[141] Shock, page 81.
[142] Bill Perkins to Susan Gordon, April 17, 1992.
[143] Ibid.
[144] Ibid.
[145] Shock, pages 80-93.

Wilson Caldwell Monk, a Master Sergeant with the all-black 320th Barrage Balloon Battalion, recalled that many of the soldiers' lessons were taught in the huge hangar and most of the balloons that were used were of the VLA variety. "It was a combination of technical and muscle power. It really wasn't complicated. Each man had a job to do." [146]

One man who served with the 317th was Loyal Whiteside, who now lives in Wildwood, Florida. Whiteside recalls that the 316th was in training at the same time as his unit "and we were in competition with them on the rifle range and other training."

During training exercises at the camp, heavy "tie-downs" would be used to anchor the balloon cables into the ground. Several of the "tie-downs" still are in the ground in farm fields where the camp formerly was located.

Whiteside recalled, "There were about 20 tie-downs in a circle. The balloon always had to be heading into the wind, since it had no frame, a strong wind could cave in the sides. We were always in contact with the weather bureau, got weather reports about every 10 minutes and it took about a dozen men to turn the balloon into the wind."

The operators "had to shift the balloon each time the wind would change a few degrees. Was not fun in a strong storm with shifting winds," Whiteside said. "We were told that was one reason the camp was in Northwest Tennessee, as the weather was always shifting." [147]

CONSTRUCTING THE BALLOON BEDS

The late Herschel K. Smith, formerly of Bruceton, was in the enviable position of being awarded the contract to build the 'tie-down' beds at the camp.

His son, Jonathan Smith of Jackson, interviewed his father before the elder Smith died in 1998, about his role in constructing the camp. "In 1941, the U.S. Army Corps of Engineers, West Memphis, Arkansas, awarded the contract to him to build the barrage balloon 'beds' at Camp Tyson for his bid of $21,000," Smith said. [148]

"He had the assistance of a reliable foreman on the job. The gravel and other materials for the work were purchased from T.L. Herbert

---

[146] Wilson Caldwell Monk, telephone interview with author, April 13, 2010.
[147] Loyal Whiteside, email communication with author, Nov. 25, 2009.
[148] Jonathan Smith letter to author, February 1, 2010.

AS IF THEY WERE OURS

and Sons, Nashville, and the cement was ordered from the Marquette Cement Co.," he said. [149]

Each balloon bed at the camp "consisted of a concrete circle that had an approximately 70 ft. diameter, in the center of which was a rectangular hole measuring about 3 ft. by 6 ft., poured as noted in concrete, in which was anchored a bolt or bolt-like device by which the large inflated balloon was held/secured. Located around the beds were 8 or 10 circular holes, 2 ft. by 2 ft. by 21/2 ft. deep, made of concrete, from which there were steel anchor rods used in the actual maneuvering of the balloon," Smith said his father told him. Smith was contracted to construct 28 barrage-balloon beds at the site. The beds were in the camp, but not visible from the main highway, he said. [150]

Training also involved launching the balloons from the backs of army trucks. According to the April 28, 1943, issue of "The Gas Bag," the camp newspaper, the soldiers at the camp enjoyed the truck training.

In an article about the Barrage Balloon School, Training Battery No. 2, Private First Class Morton Berman described it this way: "Yep! Any truck with a G.I. bumper can now be obtained complete with balloon. The Balloon Operators have done an Edison again and produced a truck mast for the little balloons. It's easy to make and simple to operate. The students are having great fun driving along at breakneck speed (25 mph max.) and skidding (simulated) to a stop at an overhead obstacle. Then they unfasten the balloon, run around the obstacle, attach the balloon again and drive on.

Passing trucks present ticklish moments, but all have survived. The right of way system has worked fine. Everyone seems to be enjoying the practice and all seem to be sold on both the idea and the equipment." [151]

As training produced more experience with the balloons, the Army determined that the VHA (Very High Altitude) balloon with a maximum altitude of 15,000 feet was not a requirement. Too many men were required and winch and cable problems were encountered. "The VHA balloon project was discontinued in May 1942 and never reopened." [152]

---

[149] Ibid.

[150] Ibid.

[151] Pfc. Morton W. Berman, "Day Room Orderly is the 'Forgotten Man' in Army Life, Says Btry Reporter," *The Gas Bag*, April 28, 1943, page 4.

[152] Shock, page 66.

The barrage balloon battalion strength was revised to 43 officers and 1,106 enlisted men when the Low Altitude (LA) balloon was adopted. It was approximately 80 feet in length, 30 feet in diameter, 23,500 cubic feet volume at its flying height of 5,000 feet, according to "The U.S. Barrage Balloon program." The LA balloons became standard for the program in 1942. [153]

In July of 1943, the training period was increased to 22 weeks from the original 10 weeks. "The first eight weeks was devoted to basic individual training, the ninth through the seventeenth weeks for unit training, the eighteenth and nineteenth weeks for tests including physical tests, balloon handling and operating and tactical exercises. The last three weeks involved advance unit training with extensive field exercises including flying of balloons and preparation for movement along with the correction of deficiencies noted in the field." [154]

As time went on, three of the battalions in training were dispatched to the West Coast and a fourth was sent to defend the Sault Ste. Marie Canal in Michigan.

STAYING IN PUP TENTS

One of those sent to the West Coast was Maj. William Jones of Tavares, Fla. Jones was a member of the 308th Coast Artillery Barrage Balloon Battalion and said, "We traveled by train to Olympia, Wash., where we left the train and were trucked to a little town of Chico, Wash., where the officers stayed in the school house and we enlisted personnel stayed in our 'pup tents' in the school yard. Bathed and shaved in a trout stream on the property." The battalion flew balloons over Bremerton Shipyard, a Marine ammunition dump and base, he said. [155]

Contracts for the barrage balloons and other equipment necessary for the school were spread around among recognizable American companies: "Barrage balloon manufacturers included Goodyear Tire and Rubber Co., B.F. Goodrich, General Tire and Rubber Co., Firestone Tire and Rubber Co., U.S. Rubber and Air Cruisers, Inc., and others.

---

[153] Ibid.
[154] Shock, page 67.
[155] Major William Jones, letter to author, June 25, 2009.

Winches were produced by the Marmon-Harrington Co., Gar-Wood Industries, James Cunningham Co., Pacific Car and Foundry Co., and the Wilson Manufacturing Co.

Balloon cable was produced by the John A. Roebling Co., American Chain and Cable Co., American Steel and Wire Co. and Bethlehem Steel Co.

Hydrogen generating and helium purification equipment was produced by the Independent Engineering Co. of O'Fallon, Ill." [156]

The first barrage balloons used in the U.S. were acquired from the British, thanks to the land-lease program, but the British and U.S. were not the only countries that used the balloons. "It has also been reported that the Russians used them for protection over Moscow. Barrage balloons were used by both the Germans and the Japanese during World War II. They were in place over Tokyo during the Doolittle raid of 18 April 1942." [157]

And World War II was not the last time the balloons were used. The last known use of a barrage balloon for defensive purposes was in 1966 during the Vietnam War when they were flown over Hanoi, North Vietnam, "and proved effective against low level bombing attack." [158]

Several surplus balloons were sold by the government to private companies after World War II and were used for advertising purposes. The Chevrolet division of General Motors purchased 26 balloons and used them for advertising, flying them over dealerships. [159]

Barrage balloons still are on display in various museums, including the U.S. Air Force Museum at Wright-Patterson Air Force Base in Dayton, Ohio, and the World War II Blimp Hangar Museum at Tillamook, Oregon.

---

[156] Shock, page. 42.
[157] Shock, page 46.
[158] Shock, page 47.
[159] Ibid.

Chapter 6

"It Just Exploded"

WHEN Camp Tyson soldiers were training with the barrage balloons — which they did every day — the skies over the camp appeared to be a veritable sea of floating balloons.

The barrage balloons were so big and the altitudes at which they hovered were so high that the balloons could be seen from points far and wide throughout Henry County.

"The sky was peppered with balloons," Mary June (Smallwood) Sinnema recalled. Her family farm was on Guthrie Rd., several miles from the camp. "The whole sky was full of them and when the sun reflected on the balloons, it was quite a sight." [160]

Carolyn Scarborough recalls as a girl sitting in her yard in Henry, a few miles from the camp, and watching the balloons with her friend, Mary. "We'd just sit there, watching them float around. I have to say when I first saw them, I was scared. Especially when we saw one of them float loose." [161]

When children in Paris got out of school in the afternoons one of their favorite activities was watching the balloons. Jeannette (Snow) McFarlin can recall the neighborhood kids gathering in her yard on Park Street in Paris to watch the afternoon show. "I remember all of us would count how many we saw at any one time and it seemed like we always had different numbers," she said. [162]

Betty Hart grew up on a farm between Henry and Como — several miles from the camp — and she can recall watching the balloons in the air from that distance when she was a girl. [163]

But the late Virgil Wall of Paris had the grand-daddy of all of the balloon-watching stories.

Wall was in class at the former Grove High School — located on the highest point in Henry County — and was watching out the window for balloons. "That was a real pastime for kids on Grove Hill. There were no trees around the hill like there are today and when you

---

[160] Mary June Sinnema, personal interview with author, January 19, 2010.
[161] Carolyn Scarborough, personal interview with author, March 18, 2010.
[162] Jeannette now McFarlin, personal interview with author, April 28, 2009.
[163] Betty Hart, telephone interview with author, February 15, 2010.

were on the south side, facing the camp, you could really see a lot of balloons," Wall said. [164]

While Wall was day-dreaming one day, looking out the window while in Mrs. Dunn's class, a severe thunderstorm blew in.

"I'll never forget it. The balloons were in the air and then the storm started," he recalled. "They had gotten the balloons down on the ground, but I could still see one in the air. Then it started lightning. The balloon got hit by the lightning and it just exploded." Wall surmised that the balloon must have been filled with hydrogen instead of helium because of the severity of the explosion. "I saw when it blew up and then it just burned to the ground. As a kid watching it, I must say it was pretty exciting." [165]

Wall said his excitement turned to sadness later, however, when he "heard from a soldier that knew the fellow that was operating the winch that was holding the balloon. He was killed when it blew and as far as I know that was the only fatality from an accident out there." [166]

---

[164] Virgil Wall, personal interview with author, May 12, 2009.
[165] Ibid.
[166] Ibid.

Chapter 7

"Shot Down Over Lake Erie"

A S enjoyable as it was for folks to watch the barrage balloons flying over the local skies every day, it was even more fun for the spectators when balloons got loose.

With any training program, mistakes can happen — that's what training is for, after all — and balloons frequently became dislodged and floated aimlessly over the countryside "and then the soldiers would take off in jeeps and follow the balloon down until the gas in the balloon finally ran out," as the late William Crosser recalled. [167]

The balloon breakaways could be dangerous, however, and in at least one case, it led to fatal results. In October 1943, a runaway balloon which investigators believed to be from Camp Tyson interfered with an American Airlines plane, causing a crash that killed ten people, including the Lieutenant Governor of Tennessee.

The plane took a nosedive into a "thickly wooded ravine 47 miles south west of Nashville" and among the victims were six passengers, including Lieutenant Governor Blan Maxwell, and four crew members. A report indicated that a runaway balloon had been spotted over a nearby field and that Camp Tyson officials had reported a missing balloon earlier. [168]

Two telegrams in the National Archives And Records Administration (NARA) spotlight the importance the runaway balloons were for the U.S. Army. And the lengths to which some would go to retrieve them.

A telegram sent to Camp Tyson notes a balloon had been spotted over Fort Monroe, Virginia. "Balloon broke loose Camp Tyson Feb. 17, 1943, drifting in northeasterly direction,. Maybe in air two-three days. Have notified BALTO Filter Center and Maryland State Police to be on lookout." [169]

---

[167] William Crosser, telephone interview with author, August 6, 2009.

[168] "Cause Of Airline Crash A Mystery," *The Palm Beach Post*, October 17, 1943, front page.

[169] NARA, RG 319, Entry 47, Box 1319, Ordinance plants, Camp Tyson (folder 0004.4), Army Intel Decimal File (1941-45).

Another telegram, sent to the camp on June 9, 1943, informs of a more dire act used to stop a wayward balloon in its tracks. "Barrage Balloon which broke loose from mooring at Camp Tyson shot down over Lake Erie returned to Akron for repairs, General Tire and Rubber Co."[170]

But for local people, the runaway balloons were a source of fascination.

"Our farm bordered the Tyson property when I was young," Ralph Anderson of Paris remembered, "and we had two or three land on our farm. The military would always come in and get them wherever they had come down."[171]

Jim Olive was raised on a farm several miles from Camp Tyson — between Cottage Grove and Como — and recalled when a runaway balloon landed on his family's farm. "I was a teenager and only one of our neighbors had a phone, but word got to the Camp that the balloon was there and the soldiers came out and got it," he said. "For me as a teenager, that was pretty exciting."[172]

RUNAWAY BALLOON AS LEARNING EXPERIENCE

Schoolchildren in the VanDyke area — also several miles from the camp — were "in awe of the huge balloons flying over," Nelda Pinson said. "We didn't know whether to be afraid or take in the unusual sights. We often speculated as to what might happen if a balloon fell. Our teacher, Mrs. Mary Evelyn Gorman, told us it was very unlikely to happen."[173]

But Mrs. Gorman was wrong, as the VanDyke schoolchildren found out, when they had the opportunity to see a downed balloon up close and personal one day. "A fourth of a mile from school, it happened, as one fell in a field," Pinson said. "I don't recall how the school got the news so quickly. Mrs. Gorman, being both teacher and principal, decided to use this as a field trip and I'm sure she intended it to be a learning experience also."[174]

---

[170] Ibid.
[171] Ralph Anderson, personal interview with author, August 6, 2009.
[172] James Olive, personal interview with author, May 7, 2009.
[173] Nelda Pinson, personal interview with author, October 9, 2009.
[174] Ibid.

As schoolchildren, Pinson said, "I'm not sure what we learned from this alien object, but we were certainly in awe. Of course, we were not allowed to go near it, but it was very pretty material of black and white. It reminded me of a hot air balloon." [175]

Pinson's father worked for the Camp Tyson Fire Department "and we, my sisters, brother and I, were proud to say that our Daddy worked at Camp Tyson, but I'm sure many other children's fathers worked there, too." The field trip sadly did not last too long, she said, "as the Army came out and got the balloon pretty soon." [176]

The late Rebecca Goins recalled that one of the runaway balloons landed on her family's property right outside of the camp one night "and it was hung over a fence. I remember my father helped them hold the balloon while they were trying to collect it." [177]

Joe Lankford, whose family farm was just a few miles from Goins' family, said the runaway balloons often caused damage for nearby property owners. "If one broke lose, it would tear down your fences," he said. [178]

Let's face it, most everyone who was in the local area during Camp Tyson days can recall seeing runaway balloons, but how many realized just how far and wide the balloons sometimes would travel? Barrage balloons were known to have landed as far afield as Kentucky, Indiana, Arkansas, and Ohio and retrieving them became quite a chore for the soldiers.

AVOIDING FATAL EXPLOSIONS

The Army was so concerned at the prospect of runaway balloons that it placed an article in *The Paris Post-Intelligencer* warning the public that such instances could occur. "In view of the fact that the Barrage Balloon Training Center at Camp Tyson is a working organization and that barrage balloons will be floating daily from the center...There is a possibility that some of the balloons may break loose from their moorings and land in the vicinity of Paris and adjoining towns and

---

[175] Ibid.
[176] Ibid.
[177] Rebecca Goins, personal interview with author, October 2, 2009.
[178] Joe Lankford, personal interview with author, April 21, 2009.

communities. They have been known to drift for hundreds of miles before coming to earth, "[179] the article read.

According to the Army, "In many cases, highly explosive gas will be used to inflate the balloons and it is against this and other features, that the public is urgently warned. To avoid a fatal explosion, all fires must be kept away from a stranded balloon, and no one is to light a match or smoke a cigar or cigarette near the balloons for the smallest spark is capable of igniting the gas and causing a terrific explosion."[180]

Equally as dangerous, according to the Army warning, were the cables which would be hanging from the runaway balloons and could come in contact with power lines. "The cables also generate electricity by dragging over the ground and can produce a shock which might be fatal...Many of the balloons will be unmarked regarding their danger and therefore, the public is warned to stay at a safe distance from all stranded balloons and balloon cables."[181]

Dudley Atkins of Alabama was a balloon rigger at Camp Tyson and he recalled when a cable snapped during training one day "and one of our balloons got away and burst. It landed 10 miles away from camp and we brought it in."[182]

The balloon "had a 15 foot tear and we sewed it up, cemented it, patched it, so we could make it fly again," Atkins said. [183]

Atkins could attest personally to the dangers inherent in the runaway balloons. "I remember in the spring of 1943, we were on Tennessee maneuvers, then on to Louisiana for maneuvers, and lightning struck a balloon. At camp, we used helium, but in the field we used hydrogen gas. You couldn't take a chance with hydrogen, it just started fire."[184]

As the Army's warning indicated, balloons could land hundreds of miles away.

"We trained every day out in the fields at the camp," the late Eddie Clericuzio of McKenzie, said, "I remember one day one of the balloons

---

[179] "Run-Away Balloons May Be Highly Dangerous," *The Paris Post-Intelligencer*, May 1, 1942.
[180] Ibid.
[181] Ibid.
[182] Dudley Atkins interview with Susan Gordon of the Tennessee Historical Society, April 17, 1992, for "Home Front" project.
[183] Ibid.
[184] Ibid.

AS IF THEY WERE OURS

got loose and landed in some lady's yard in Kentucky. We had to send some trucks out there to go get it." [185]

The late Bill Perkins of Paris, who operated the winch for the barrage balloons, said, "Several of the balloons got loose and they'd wind up breaking. We'd find them all the way over to Humboldt, Van Dyke, Puryear, Whitlock." [186]

BREAKAWAYS LANDING ALL OVER

The 302nd Coast Artillery Barrage Balloon Battalion, the outfit which launched the first barrage balloon at Camp Tyson, did lose several balloons in its training. In April of 1942, the unit lost two balloons "due to break-away moorings. One balloon was retrieved at Clay, Kentucky, and the second balloon at Bruceton, Tennessee. Another balloon was lost and recovered at Minor Hill, Tennessee, in May 1942." [187]

Balloon breakaways continued to plague the Camp during training, with balloons being retrieved in Proctor, North Carolina; Camp Campbell, Kentucky; Darden, Tennessee; Spring Creek, Tennessee; and as far as Frenchmans Bayou, Arkansas, in October, 1943, according to research by James Shock, author of "The U.S. Army Barrage Balloon Program." [188]

A *Paris Post-Intelligencer* article from May 28, 1942, reported that an escaped balloon was found near Pulaski, Alabama. "The balloon was only slightly damaged when found, and has already been returned to the local balloon training center by a retrieving crew."

Former soldier Loyal Whiteside of Florida kept an undated newspaper article in his scrapbook which reported a runaway balloon had landed near New Corydon, Indiana. According to the article, a retrieving detail of 25 soldiers was sent to recover the balloon. The soldiers drove all night in the truck and found the balloon hanging on a tree with all the hydrogen out of it. The soldiers wound the cable up, ate breakfast in the field and started back for camp. [189]

---

[185] Eddie Clericuzio, personal interview with author, May 11, 2009.

[186] Bill Perkins to Susan Gordon for The Tennessee Historical Society 'Home Front Project," April 17, 1992.

[187] *The U.S. Army Barrage Balloon Program*, page 81.

[188] Ibid.

[189] "Help Nab Runaway Balloon," undated newspaper article, property of Loyal Whiteside.

The Camp Tyson newspaper, *The Gas Bag*, had some fun in its reportage of a lost balloon in its April 28, 1943, edition. In an article written by Pvt. Jack Redmond, it was reported that Battery A of the 316th had lost its first balloon. A crew, with equipment and rations packed, was given instructions to retrieve the balloon at Fairview, Tennessee." [190]

On the way there, the crew stopped at the middle Tennessee town of Dickson for directions and were told the balloon had been sighted at Primm Springs, which was nowhere near Fairview. Pvt. Redmond noted that the crew was redirected "to the hills of Tennessee. And when we say hills, we mean HILLS!" [191]

"The trucks then set off and traveled up hills, down dales, through streams, cow pastures and what have you," Pvt. Redmond wrote. "To make matters worse, the roads at times were blocked by farm cattle." [192]

The balloon was finally spotted lying on top of a hill, "above a clearing, past a stream, and near the back of a farmhouse," which required chopping down trees and brush. The soldiers finally were able to grab the balloon, spread it on the ground and deflate the rest of the gas from it before placing it in a truck." [193]

Dusk was falling and "rather than mess up the victory and take the men over the rough roads," Redmond wrote, the crew was directed to Dickson, where they ate their first meal — "chow and coffee" — in nine hours. "The trucks returned to camp at 1:45 a.m., exactly twelve hours after they had left." [194]

---

[190] Pvt. Jack Redmond, "Story of First Lost Balloon and Task of Retrievers Fascinates 316th BN Writer," *The Gas Bag*, April 28, 1943, page 8.
[191] Ibid.
[192] Ibid.
[193] Ibid.
[194] Ibid.

Chapter 8

Good Times For Entrepreneurs

THE dormant local economy began to take off during the construction phase of Camp Tyson and grew in leaps and bounds during its operation. Suddenly, jobs were available for everyone that wanted them. Local businesses that had been in decline began to pick up and the opening of the camp produced side businesses for energetic entrepreneurs — and that included men, women, and, even children.

Even in a time of great national sacrifice, these were boom times for Henry County and its population rested easier knowing that their bills could finally be paid on time.

Local households may have been overcrowded and uncomfortable from the great numbers of construction workers, soldiers and their families staying with them, but it also meant rent was being collected from the new guests. The extra money allowed people to relax and take a deep breath for the first time in a long time.

"Workmen by the thousands came to Paris," according to Bryant Williams, whose family owned *The Paris Post-Intelligencer*. "They spent their money in Paris and it really helped our economy." [195]

With the opening of the camp came even more opportunities for those wise and creative enough to take advantage and those opportunities came in a wide variety from buses, sandwiches, and pop machines. Even bootlegged liquor.

ESTABLISHING THE BUS LINE

One of the first to see a growing need and act on it was Frank Blake, who later became mayor of Paris. He knew several things: an army camp was just down the road, gas was rationed, and there was no public transportation.

He applied for and was awarded a government contract to establish a bus line from Paris to the Camp at Routon, seven miles distance.

---

[195] Bryant Williams, personal interview with author, April 7, 2006.

The new bus line opened February 20, 1942, shortly after the camp was open for business. Blake had operated a tire store at the corner of Brewer and Wood Streets, which he closed out, converting the building into a bus depot. He also already operated a transport terminal on N. Market Street, which was expanded to provide a service center for the buses.

The buses operated on 30-minute schedules, with a fare of 15 cents for one-way trip and 25 cents for round trips. Books were issued for $2, which contained tickets for 20 trips.

Operating 24 hours a day, the line initially operated with two buses. The 30-minute schedule was in force from 5 a.m. to midnight, with the second shift of midnight to 5 a.m. operating every two hours.

Eventually more buses were added, with bus stops located throughout Paris. The routes carried the 30-passenger buses from Paris to Gate No. 1 at the camp, with exits from Gate No. 2 for the return trip to Paris. [196]

The buses were a popular — and necessary — mode of travel, but also could be a bit hazardous.

James Wilson of Paris, who served at the Camp in its medical detachment, recalls that the buses provided weary — and a bit boozy — soldiers with a way to get to camp after nights on the town. "They'd put 30-40 soldiers in there and every time they'd put on the brakes, everybody would go forward and crash into each other. A lot of guys got sick that way." [197]

Where there are soldiers, there is a need for food and plenty of local people were ready and willing to fulfill that need.

MAKING 125 SANDWICHES A DAY

Julia Hutson supplied meals for soldiers and their families from her home at the corner of Edgewood and Park Streets in Paris. "We had this big house and men were stopping here asking for rooms. We rented the attic, the back and front rooms and then they wanted to know if I'd make their lunches. Then three men who were staying up the street asked, so I was making lunches for nine men altogether." [198]

---

[196] "Paris To Camp Bus Line Will Be Started On February 20," *The Paris Post-Intelligencer*, February 3, 1942.

[197] James Wilson, personal interview with author, April 16, 2009.

[198] Julia Hutson, personal interview, May 10, 2009.

Word got out and shortly the manager at the L&N Railroad called her, asking if she could supply lunches for the soldiers passing through on the troop trains. That would mean anywhere from 50-125 lunches. Hutson agreed.

On a regular basis, she made sack lunches at her home and carried them to the depot on the west side of Paris. "I would make two sandwiches, a dessert and fruit. I'd make ham or pimiento or cheese or bologna for the sandwiches, put in two cupcakes or cookies and some fruit." [199]

Her deliveries were usually made on weekdays and Saturdays, she said, "but I remember one Sunday we were getting ready for church and they called and said there was a troop train coming and they needed 90 lunches. I didn't have anything ready, so I called John B. Arnett (owner of Arnett's Grocery) and he opened up the store for me." [200]

Rachel's Café was located in a prime spot to take advantage of the camp. "The USO was right across the street from us" on E. Washington St., according to Fern Pierce, whose family operated the café. Pierce and her three sisters went into the restaurant business, which their father purchased for them. Their specialties were "good home-cooking, plate lunches for 35 cents or 25 cents" which included meat, two vegetables, bread and a drink, she said. [201]

"We had one of the best hamburgers around. Our cook made his own relish and we had the best fried pies," Pierce said. Willie Neese, who also later became mayor of Paris, installed a sound system at the USO, which made it even more popular. "Soldiers would come into town wanting to go somewhere" and Rachel's café was in a prime spot to attract them. "It was boom times for everyone," she said. [202]

DELIVERIES WERE FUN

Local grocery stores were busier than ever, including Schofner and Thompson's store on E. Washington St. "I was a little fella, but I can remember we took truck loads of groceries to the camp," Franklyn Thompson recalled. "I can remember us delivering groceries to the of-

---

[199] Ibid.

[200] Shannon McFarlin, "From People To Plant: A Lifetime Of Caregiving," *The Paris Post-Intelligencer*, September 2004, page 3B.

[201] Fern Pierce to Susan Gordon of The Tennessee Historical Society for its "Home Front" project, April 17, 1992.

[202] Ibid.

ficers' club. We'd go out there regularly. The camp had a great impact on us and on the town." [203]

The family of Virginia Sinclair operated Timmons' men's store, which was five miles from the camp. The store was owned by Homer Timmons and sold men's clothing.

Sinclair, who now lives in Memphis, recalls how busy the store was when the camp opened. "We sold men's clothing for years and when the camp opened, they sold anything military. Anything that the boys could buy. Clothing with bars and insignia on them." [204]

Sinclair's mother, Cecil Atkins, did sewing and alterations for the store. "She worked there in exchange for rent and I was in high school, so naturally I was in there all the time. I remember the camp boys would come there every Saturday morning by the busloads. They'd buy things and then just sit around and talk. It was a fun time." The soldiers "came from everywhere," she said. "I can still remember the first names of a few." Some of the soldiers asked her out, which also was fun. "I remember going to the show with a couple of them. I don't really know when I had time to do my studying." [205]

The Lasater family owned the Coca-Cola bottling plant in Paris and the boon in the economy "greatly enhanced the business," Jack Lasater said. Lasater's father, J.P. Lasater was mayor during the Camp Tyson days and operated the plant, as well. "We had to add more shifts at the plant," Lasater said. "Commercially, Camp Tyson was a great help to Paris. The milk and bread companies did more business, our company did more business. Everyone benefited." [206]

Bootleggers also benefited, with William Crosser and Roland Atkinson recalling a great deal of activity in that realm occurring while the camp was in operation. (See Chapters 11 and 14) "There was a lot of night clubbing going on then," Fern Pierce said. "We had two theaters in town then and more nightclubs in Paris than there are now." [207]

As a civilian employee with the motor pool, the late Charles "Buddy" Porter was in a good position to witness much of the bootlegging activities there. His nephew, Jeff Wygul, said his uncle loved to regale family members with stories about working at the camp. "He

---

[203] Franklyn Thompson personal interview with author,February 12, 2004.
[204] Virginia Atkins Sinclair, telephone interview with author, October 1, 2009.
[205] Ibid.
[206] Jack Lasater, telephone interview with author, February 15, 2010.
[207] Pierce to Susan Gordon.

said bootlegging was a major money-maker out there. He said soldiers would use General Maynard's car to bring the moonshine into the camp at night. Of course, the general didn't know his car was being used for that. But nobody is going to search the general's car, so they could take it in to town at night, saying it was for something official and then go get the moonshine and bring it back into camp without anyone looking inside it to see what was really there." [208]

And it wasn't just adults who were benefiting.

Dorothy Vaughn remembers that soldiers used to pay her brother "to shine their shoes, even when they didn't really need it." [209]

When he was a teenager, Virgil Wall said he delivered newspapers. "I used to try to get to the L&N depot when the troop trains would come in. The train would be crammed full of soldiers and they would do anything for a break. I could sell a lot of papers to them." [210]

EVERYBODY NEEDS ICE

Another business that boomed was the People's Ice and Coal Co., in Paris, which found itself filling all the camp's ice needs. The company was operated by C.W. Richardson on Porter Street and during the war, faced with a manpower shortage, their son, Jerry, found a steady job as a delivery person.

Richardson's widow, Gene, recalled that her husband used to talk with humor of the deliveries he made to the camp, especially since he was too young to have a driver's license at the time.

"He said one time he was making a routine trip to the camp to deliver ice and he came to the checkpoint and someone asked to see his license. It wasn't the person who normally worked at the checkpoint. Jerry said, 'I don't have a license' and the soldier told him he couldn't come in without one. Jerry got aggravated and he said, 'Look, I got the ice here and if you want it, you're going to have to let me pass because I'm the only one who can deliver it to you.' He said the soldier thought for a second and then said, 'Well, go on in then.'" [211]

---

[208] Jeff Wygul, personal interview with author, November 12, 2010.

[209] Dorothy Vaughn, telephone interview with author, June 9, 2009.

[210] Virgil Wall personal interview, May 12, 2009.

[211] Gene Richardson, telephone interview with author, December 7, 2010.

Dan Nealon of Paris was an especially wily young man. He lived in the house next to the Barton mansion on N. Poplar St. (which now is the Paris-Henry County Heritage Center). The huge Barton home became headquarters for the contractors who were building the camp.

"The contractors were all over the building, upstairs and downstairs. The Bartons weren't living there then and they leased it to the contractors. This would be late spring of '41." Nealon was in 8th grade and began managing the Coca-Cola machine that had been installed in the building. "I remember Ven-Dall made the machine and Coca-Cola was the provider and I rented it. I guess I was an early entrepreneur. I put it right near the back entrance, the N. Market Street entrance, just three steps under the landing. A good spot for it." [212]

Nealon said, "I made so much money with that machine that the contractors decided they'd manage it for themselves. But I provided a real good service for two full months. Right until it was time for me to go back to school." [213]

---

[212] Dan Nealon, telephone interview with author, May 14, 2009.
[213] Ibid.

Chapter 9

"They Would Throw Hands Of Candy And Gum At Us As They Marched By"

BARRAGE balloon training was not the only activity that took up the Camp Tyson soldiers' time.

As with any other Army camp, soldiers were expected to stay in shape and what better way than with regular hikes? At Camp Tyson, the hikes could be 15 or more miles long and took the soldiers far and wide through the countryside surrounding the camp.

With soldiers carrying their complete gear and weapons and with the hikes occurring through the back-country hills of Tennessee in all types of weather: from hot and humid to rain and snow, the hikes were quite arduous.

The soldiers would encounter much that probably seemed rather foreign to many of them during the hikes: from large snakes to talkative farmers. And for the lengthier maneuvers which took the Camp Tyson soldiers on overnight and sometimes week-long military exercises to Mississippi, Louisiana, and the even hillier sections of Tennessee, a soldier's life could be extremely weary. But one soldier recalled having his first encounter with "white lightning" during an overnight maneuver, which probably helped ease the pain.

The hikes were a source of wonder for Henry Countians who viewed the soldiers on their regular jaunts through the countryside.

"SEEMED SO OMINOUS"

When she was a child, Ruffalene Webb of Henry lived in a former stagecoach station on Macedonia Rd. and she can vividly recall her reaction to the hikes. "Those troops would march by our house and it would just scare me to death. You could hear them coming because they marched so loud, but because of the hill and the curve right before our house, we couldn't see them until they were right in front of the house. It just seemed so ominous to me as a child." [214]

Her father would be at work, she said, "and Mother would reassure me, or try to. It was always during the day that they came by our house and there were always three groups of them that went by. Mother told me to wave as they went by. But I just remember being scared to death of them." [215]

Lou Carter of Paris remembers soldiers "going on extended hikes" past their home near Mansfield, several miles away from the camp. "We used to see them doing their military drills way out there," he said. [216]

For Nelda Pinson of Paris, watching the soldiers marching was a pleasant memory. Pinson went to school in the small community of VanDyke, several miles from the camp, "and they would march from Camp Tyson through VanDyke and back to camp, always in unison. They always looked so nice in their uniforms, all starched. Today our military is mostly seen in camouflage dress and I know they're equally as good soldiers, but those nice starched uniforms and hats would make you proud." [217]

Pinson recalled the students would be allowed "to line up across the front of the school on the bank and they would throw hands of candy and gum at us as they marched by. This was a real treat for us children. They were like stately robots, holding their rifles on their shoulders. As a child, we thought of them as the tin soldiers." [218]

Soldiers marched by the home of Glenda McNutt of Paris, who lived on VanDyke-Routon Rd. when she was a child. "It was a regular occurrence," she said. "I remember we stood on the porch and watched. I was fascinated, not frightened. I can remember it vividly." [219]

The late Bettye Hill, who grew up in the small town of Henry near the camp, said seeing the soldiers doing maneuvers was a common sight, but also a frustrating one. "I remember you couldn't get out on

---

[214] Ruffalene Webb, personal interview with author, September 3, 2009.
[215] Ibid.
[216] Lou Carter, personal interview with author, May 28, 2009.
[217] Nelda Pinson, personal interview with author, October 9, 2009.
[218] Ibid.
[219] Glenda McNutt, personal interview with author, October 6, 2010.

the road for two or three hours sometimes because there were so many soldiers out on the roads." [220]

Dorothy Cook's family grew up near the small town of Como, several miles from the Camp. "Mother told me when the soldiers came by, people would hand them water to drink to help them along." [221]

OLD MAPS DON'T HELP

Captain James E. Lewis, Jr. of Missouri, recalled an especially interesting march experience in a personal journal he wrote about his Camp Tyson days. "I was called to headquarters and put in charge of a 10 mile march. I was given the march route and the time...A sergeant and I took a jeep and went out to do a recon on the march route. To my horror, we found the march followed an old map of the post. We would follow a road for one or two miles and find a fence crossing the road. The only way to cover the area was to march across fields and pick up the roads beyond the fences. It was a horrible mess." [222]

With no time to make any changes, Lewis said he assembled the men and officers, noting that the nurses had a separate march. "We marched up the road for about ½ mile when I noticed the commanding general's staff car had parked at the side of the road and one of our officers was pointing at me. I ran forward and reported. The General was red faced and furious. He asked me what my IP (initial point) was. I replied 1300. He stated very strongly that he did not think I was going to make it. I agreed. One doesn't argue with generals. However, I thought the IP was where we began our march. I discovered the point of the march we were supposed to reach by 1300 was still a quarter of a mile away! This was the beginning of a very bad afternoon." [223]

Lewis said as the soldiers marched across fields and around the obstructions, General Maynard was watching from his parked vehicle. "I must have saluted him a dozen times and his face was even more red each time." [224]

---

[220] Bettye Hill, personal interview with author, March 30, 2010.

[221] Dorothy Cook, personal interview with author, March 26, 2010.

[222] "A Personal Journal of World War II, 1941-1946, Capt. James E. Lewis, Jr., MCAUS, 0475690" by James E. Lewis, Jr., MD, 1995, on Rocket League web site, DadWWII.

[223] Ibid.

[224] Ibid.

The Gas Bag, the Camp Tyson newspaper, used to devote quite a bit of ink to the hikes in its regular editions, usually with humor.

"The recent 14-mile hike proved that members of the Cadre Pool are not only good-weather hikers, but are excellent 'mudders' as well and can travel through that Tennessee chocolate with the agility of a two-year-old pacer," *The Gas Bag* reported in its April 28, 1943 issue. [225]

A newlywed bridegroom, Pvt. Fred Mariotti, was "conspicuous in his absence" during that hike, the paper reported, "but then you can't expect one to hike through that mud after wading the Sea of Matrimony. However, he has graciously consented to do a special stretch of latrine duty in lieu of the hike he so absently-mindedly overlooked." [226]

Pvt. Samuel Unger of the 102nd Battery, First Platoon, noted that "Hikes are boring enough but you don't know what boring is until you walk alongside a farmer who talks of nothing but his litter of pigs for four hours." [227]

In his column for *The Gas Bag*, Pvt. Unger also reported a too-close encounter with a reptile during a recent hike. "When a passerby noticed J.D. Baker shouting 'at ease, at ease' and beating something on the ground, he stopped and looked on in amazement. Not until the passerby was absolutely satisfied that J.D. wasn't crazy, did he leave. Even then, it was with a smile on his face. What actually happened was that Pfc. Baker had to get the snake before the snake got Baker." [228]

Former Soldier Loyal Whiteside of Wildwood, Fla., was a member of the 317th Barrage Balloon Battalion. "We were always taking hikes, even many at night. 20-25 miles was not uncommon and all the roads were dirt outside the main camp area. The road names were not known by us, but many of the balloon sites were on old farms and we would use the houses as cooking and sleeping area for the crews." [229]

Whiteside has kept his scrapbooks from his Camp Tyson days, which include several photos from overnight maneuvers that his pla-

---

[225] "Enlisted Cadre Pool Is Rapidly Becoming A Lake," *The Gas Bag*, April 28, 1943, page 9.

[226] Ibid.

[227] Pvt. Samuel Unger, "There's Always Something Interesting Going On In VLA Batteries." The Gas Bag, April 28, 1943, page 6.

[228] Ibid.

[229] Loyal Whiteside, personal interview Nov. 25, 2009.

toon took to Camp Forrest, near Memphis. The overnight trips were a source of lengthy articles in *The Gas Bag*, as well.

MAKING DO WITH FOOTBALL AND DANCING

"After our great overnight convoy last week, the boys are now running around calling themselves real honest-to-goodness outdoorsmen — and soldiers, too," Staff Sgt. Max Lipsky of the Headquarters Battery wrote of a trip to Natchez Trace Park near Memphis in April of 1943. [230]

"We landed at Natchez Trace Park and with instructions to pitch tents so they would not be seen by enemy planes," Lipsky wrote. The soldiers' abilities on that score met with mixed reviews, with several unable to find their own tent when nightfall came. [231]

The soldiers' sleep was interrupted with rainfall and cold, he said, but before that, they were entertained with a football game on the grounds. The trip wasn't without pleasure, since a dance was held in the lodge at the park and "girls from Lexington were convoyed in and a good time was had by all, particularly during the jitterbug contest. The old floor was really rockin' while this was being staged." [232]

The late Dudley Atkins of Alabama also recalled the arduous nature of the overnight maneuvers. "I remember we went to Lebanon Carthage, Murphreesboro, Cookeville, in Tennessee, so that we could protect the bridge over the Cumberland. Well, actually, we were protecting the river and rail yards, things of that nature. We were gone for two months, from April through June of 1943, doing that." [233]

But Atkins was able to learn something new on those maneuvers. "I learned how to drink white lightning on those trips," he said. "I can remember staggering through the chow line many a time." [234]

---

[230] Staff Sergeant Max Lipsky, "Convoy To Natchez Trace Park Brings Men In 317th Battalion A Real Thrill," *The Gas Bag*, April 28, 1943, page 10.

[231] Ibid.

[232] Ibid.

[233] Dudley Atkins to Susan Gordon of The Tennessee Historical Society for "Home Front" project, April 17, 1992.

[234] Ibid.

Chapter 10

"We Reached Out To Those Boys As If They Were Ours"

FOR young women in Henry County, Camp Tyson was a source of great excitement.

It was a time when good-looking soldiers in crisp uniforms flooded the streets of Paris. Their small town was suddenly much bigger and their isolated and insulated lifestyle was expanding. Soldiers from all over the country — with strange accents, cultures, and backgrounds — were right at their door steps.

Opportunities for friendships and romance with men who they ordinarily would have never met in their lifetimes were just around every corner. The war loomed large over their lives — rationing, air raid drills, and blackout curtains were a part of daily life. But Camp Tyson was right in their backyard and that meant Southern households did what they had done for generations: welcomed newcomers with all the hospitality they could muster.

Looking back on those times today, two ladies who enjoyed the social whirl of the Camp Tyson days recall them now with bittersweet affection. Active in the "Liberty Belles," Carol Morris Bowden and Ada Margaret Humphreys Atkins remember the excitement and the sadness associated with the dances that brought local girls and soldiers together in the community and the camp. They were classmates at Grove High School during the Camp Tyson days, both in their mid-teens. Bowden now lives in Montgomery, Alabama, and Atkins lives in Pensacola, Florida.

"IT FELT FALSE"

"Camp Tyson brought big changes to Paris," Bowden remembered. "Paris was the most ideal place to grow up. There was a sweet innocence about the place. For us to all of a sudden be surrounded by people from other areas...Well, we all changed." People in Paris "became more cautious. But at the same time, every boy we saw in uniform reminded us of our boys in uniform," she said. "And we wanted to make them feel at home." [235]

Bowden remembers that the girls of Henry County "were just delighted to have these soldiers around. The boys were homesick and they enjoyed coming to our houses, listening to records, talking. We introduced them to Southern cooking. But, we felt so guilty so much for having a good time." Her brother went to military school for a year and part of his uniform was a Navy pea jacket. "I remember I loved to wear that, but after Camp Tyson arrived, I wasn't comfortable doing that anymore. It felt false." [236]

Several young men from Paris, ones who Bowden knew well, died overseas during World War II "and things really hit home for us then. I think that was why Paris reached out to those boys as if they were ours," she said. [237]

"All in all, for people my age, it was very exciting, even while it was heart-breaking, to go to all the dances that were a part of our social life during the Camp Tyson days," she said. [238]

There were plenty of places where young people could mingle locally during the war years. The City Auditorium in Paris was a popular venue for dances, as were the USO clubs for both black and white soldiers (for white soldiers, the USO was located on the second floor of the former Paris Board of Utilities Building on W. Washington St. and for black soldiers, the USO was on Rison St. in a still-existing large white auditorium-sized building).

But one of the most popular spots was the Service Club at the camp.

The Paris Post-Intelligencer published a 75th anniversary edition in the spring of 1941 and dedicated it to Camp Tyson. The special edition was full of articles and photos of the new camp. "Located directly opposite the Camp Theater near the heart of the camp area," one of the articles read, "the Service Club outwardly is unpretentious in appearance, but once inside, the visitor is deeply impressed at the extent to

---

[235] Carol Morris Bowden, telephone interview with author, March 10, 2010.
[236] Ibid.
[237] Ibid.
[238] Ibid.

which the U.S. War Department has gone to keep the morale of the soldier at that level so necessary in the winning of wars." [239]

According to the article, the Service Club was "attractive in arrangement and design" and its interior centered around a huge hall which served as a dance floor and auditorium for dances, band concerts, plays and the like. "At other times, it is used as a large lounge furnished with comfortable chairs, writing tables and musical instruments for the benefit of the soldiers." [240]

As editor emeritus of *The Paris Post-Intelligencer*, the late Bryant Williams wrote weekly columns for the paper, which he called "Post-mortems" and which were later compiled into three books. According to Williams, the Service Club was under the supervision of Lt. Col. Donald M. Sensing, "a regular army officer of many years' experience, and who was praised for the efficient manner in which he filled the office of Special Service Officer." [241]

Sensing was assisted by two ladies, Mary K. Landis of Nashville, senior hostess, and Murrell Wright of Fayetteville, junior hostess, according to Williams. [242] Their photographs, along with other interior shots of the well-appointed Service Club are among those contained throughout the Camp Tyson yearbook. [243]

The Service Club offered many services to the camp's enlisted men, including a library and reading room, a cafeteria and soda fountain room, a grand piano and juke box. "A long, screened porch distinguished the Service Club from its neighboring building, and doors led into the foyer where cloak rooms were located. Large double doors opened into the main hall, which was finished in natural color and rustic design. The hall was two stories high, with a mezzanine extending completely around it." [244]

Bowden recalled the club with vivid detail, since she and her friends were regular attendees at the Saturday night dances there. "I

---

[239] Bryant Williams, "Many Locals Have Fond Memories of Service Club Out At Camp Tyson," Post-mortems column, *The Paris Post-Intelligencer*, October 4, 1993.

[240] Ibid.

[241] Ibid.

[242] Ibid.

[243] *Historical and Pictorial Review, Barrage Balloon School Station Complement, Barrage Balloon Training Center*, The Army and Navy Publishing Co., Nashville, 1942.

[244] Ibid.

kept a diary my junior and senior years in high school and there's a lot in there about those dances. Who I met, what people did, etc." [245]

ALL IN GOOD TASTE

The dances were chaperoned affairs and "in good taste. Our mothers went with us. They drove us there and never left the building," Bowden said. "There was no rowdiness and it was just good, clean fun." Bowden recalled the balcony that spanned the huge hall. "The boys from Paris would sit in the balcony and watch us dancing with the soldiers," she said. [246]

"I remember there were two ladies in charge of the club and one was a younger lady, a Miss Wright, and she was the one who knew the boys and knew which ones were homesick. She would discretely ask us to spend a little extra time with the ones who need a little extra help," she said. At the beginning of each dance, Bowden said, "The soldiers would form a line and the girls would form another line and then you'd meet in the middle and dance. That would help the shy ones so they didn't have to ask someone to dance." [247]

AULD LANG SINE

The saddest parts of the dances, she said, were when the soldiers would be called for overseas duty. "I remember all of a sudden the band would play 'Auld Lang Sine' and that was a signal that the soldiers were leaving. You would dance to that final song and then the soldiers would file out one by one. You never knew where they were going or why." [248]

Camp Tyson, according to Margaret Humphreys Atkins, "had asked for girls that would come out to the camp to the dances to help with morale. Our mothers took us and we followed the rules. We couldn't leave the building at all. You could go to the snack bar for a coke, but you couldn't leave the building." That rule was an easy one to follow, "because it was a camp rule, but it was also my mother's." [249]

---

[245] Bowden interview.

[246] Ibid.

[247] Ibid.

[248] Ibid.

[249] Ibid.

Huge crowds were the norm for the dances, with girls coming from all around, not just Paris.

The camp band, under the direction of the late Tom Lonardo, would play at the dances "and I remember some of the girls dated the boys in the band. There were many romances that started in that building. Some of those romances led to marriage, too."[250]

"It was such a fun time, but it was a sad time, too," Atkins said. "You'd get to know these boys and then they'd ship out and you'd always wonder how they were doing and what had happened to them."[251]

"GOOSE-PIMPLES" ON LEAVING CAMP TYSON

Captain James E. Lewis of Missouri provided insight into what became of the soldiers as he recalled being shipped out from Camp Tyson. On January 10, 1944, the men in his outfit were abruptly notified they would be leaving camp. President Franklin D. Roosevelt had announced plans for the build-up in England and attack on Dover. Hearts among the soldiers were heavy.

His unit assembled and marched up a hill lined by barracks. "Men were sitting out watching us move up the road. Just as we came to the top of the hill, the camp band came from retreat, assembled and began to play. At the bottom of the hill, we saw the train waiting. I shall always recall this moment with 'goose pimples.' The sun was a deep red, setting in the west. We moved into single file and stepped on the train platform. As each Pullman car filled the train moved slowly forward. As the last man entered the train, the train was moving toward the main line. It was a beautifully planned maneuver and very exciting."[252]

---

[250] Ibid.

[251] Ibid.

[252] "A Personal Journal of World War II, 1941-1946" by Captain James E. Lewis, Jr., op.cit.

As If They Were Ours

Chapter 11

War Is Hell,
But Pretty Entertaining, Too

MAINTAINING morale was an essential task for the military during World War II and, despite its remote and rural location, Camp Tyson provided a steady diet of entertainment for its soldiers. Whether it be movies, dances, or having a close brush with big-time movie stars, the soldiers had plenty to keep them occupied during their off-hours.

And, if they were looking to forget they were in the Army for a few hours, there was always some good ole' Southern 'white lightning' to be purchased in the woods surrounding the camp.

Jeanette MacDonald was not the only big-name movie star to grace the stage at Camp Tyson. (See Chapter 12). Eddie Bracken — described as "one of movie land's top notch comedians" in the camp newspaper — brought his entire USO show to the camp. By the time of Bracken's performance in April of 1943, the Camp Amphitheater had been enlarged to seat 7,000. All of those seats were available for Bracken's show, for which officers and enlisted men were allowed to invite guests. [253]

Bracken's show was quite the production, with performers including Maestro Bardo, "gifted with a line of spontaneous patter"; the vaudeville duo of Lewis and Ames; a "talented mimic" Wally West; the comedy-dancing act of Chester Fredericks and Kay Wilson; and the Tip Top Girls, described as "six lovely and talented acrobatic dancers who perform... in a number of fast-paced routines." Providing a musical backdrop for the show was Bill Bardo and his orchestra of fourteen musicians, which had appeared in several movies, including "The Goldwyn Follies." [254]

---

[253] "Eddie Bracken To Appear In Person At USO-Camp Show," *The Gas Bag*, April 28, 1943, front page.
[254] Ibid.

Even when Hollywood stars were not there, the soldiers of Camp Tyson were never without top-notch musical entertainment. They were fortunate to have talented musicians in the camp band, as directed by the late Tom Lonardo, a soldier who settled in Paris and went on to make a career in music.

From Providence, Rhode Island, and drafted in 1942, Lonardo said he "found myself in Camp Tyson," and learned that he had been sent to the camp "because they needed a fella who could play string bass and bass horn" for the band. [255]

Lonardo said he began forming a band once he arrived at the camp. "I took six guys who could play woodwinds, I found another guy who could play bugle" and rounded out the rest of the band. "We played all the parades, all the reviews, all the dances and we accompanied most of the shows that came to the Camp," Lonardo said, including the Jeanette MacDonald appearance. Of MacDonald, Lonardo recalled, "She had a beautiful voice."

But the camp band did not just stay on the camp. "We played off base at the USO club in Paris. We took the entire band to concerts on the court square in Paris. We even went to Big Sandy (Tennessee) to play at a town dance there," Lonardo said. "We toured all the TVA dams. Some were just being built right then. The purpose of that was for a bond tour. That tour took up a month and a half." [256]

The camp band also played an "infantry show," touring through Virginia, playing seven days a week for "an entire month. It was called 'Here's Your Infantry Show' and it was set up in football fields along the way," Lonardo said. "It was quite a show, but it was quite a chore. Two shows a day. But they were successful bond shows." [257]

Lonardo met and married a Paris girl, Claire Taylor, and his name became synonymous with music for local folks for many years. After the war, he owned and operated a popular music store in Paris for decades and continued to tour the region with the Tom Lonardo Band which he had formed when he settled into civilian life.

"My Aunt Opal (Reed Logan) worked at the Camp, in the finances section, and one of my best memories is of going to the camp to visit

---

[255] Tom Lonardo, in interview with Susan Gordon for Tennessee Historical Society "Home Front" project, April 17, 1992.

[256] Ibid.

[257] Ibid.

her," Bobbie Jean Freeland of Paris recalled. "My best memory of all is listening to Tom Lonardo's band playing, 'Stardust,' which was my favorite song." [258]

BASEBALL, MOVIES AND FLORA-DORA GIRLS

The warm months were especially pleasant for soldiers, with baseball teams organized to play both intra-league and with teams from nearby towns. The Camp's baseball team was called, naturally, "The Barrage Balloons." Classical music was on tap — in a fashion — when Chopin's "Waltz in E Sharp Minor" was performed "by the nimble-fingered Corporal Paul Kiesenwetter," as hairy-legged soldiers dancing as the "Flora-Dora Girls" cavorted behind him. [259]

The "Tyson Follies" was a venue for home-grown talent from the camp, such as the ever-present "Flora-Dora Girls," and was a time for some off-color entertainment for the troops. Soldier Ronnie Washburn wowed the soldiers in August of 1942, with a "Balloon Dance," which *The Gas Bag* reporter described as an updated version of the 'bubble dance,' in which Washburn steadily lost balloons "until, at the very last, he was well nigh destitute!" [260]

An "Activities Calendar" in the April 28, 1943, issue of the camp newspaper, *The Gas Bag*, lists a surprising number of choices for that week. Movies, such as "King of The Cowboys" with Roy Rogers, and "Edge Of Darkness" with Errol Flynn, were shown every night. Band shows were featured at various locations on base and at the Paris USO, as were talent shows, checkers tournaments, formal and informal dances, bridge nights and a 'minstrel show' performed by the black soldiers of the 320th Barrage Balloon Battalion. [261]

For the religious-minded, there was plenty to offer, with chapels for Catholic and Protestant persuasions providing numerous worship services every Sunday. For those wanting quiet time, the library reading room played classical and semi-classical recordings. The First Methodist Church in Paris also invited soldiers to a "soldiers' lounge" with light refreshments in the afternoon. [262]

---

[258] Bobbie Jean Freeland, personal interview, August 18, 2009.

[259] Sergeant John W. Morgan, "Tyson Follies Presented At Service Club" *The Gas Bag*, August 19, 1942, front page.

[260] Ibid.

[261] "Activities Calendar," *The Gas Bag*, April 28, 1943, page twelve.

[262] Ibid.

"We had quite a social life on the camp," Howard Koenen of Murray, Kentucky, remembered. "We had a lot of big occasions there. Actors and actresses from Hollywood would come there to perform. A lot of entertainment was going on there." [263]

BALLOONATICS OF '43. TOWN STUFF

Loyal Whiteside was a member of the 317th Coast Artillery Barrage Balloon Battalion who now lives in Florida. Whiteside has scrapbooks full of programs from the various musicals and plays which were produced by soldiers at the camp.

Among his collection is the program for "Balloonatics of '43," which features a drawing of a muscular male soldier in a woman's wig on its cover. The production was put on at the Post Theater at 8:30 p.m. Monday, June 28, 1943, and was written and directed by 2nd Lt. Albert Hutler and Staff Sgt. Max Lipsky, with music by 2nd Lt. Francis K. Shuman and Lyrics by Lipsky. It was quite a production, requiring a stage manager, dance director (Miss Murrel Wright of Paris), scenic designer, costumer and art designer. The Camp Tyson band, along with the 317th musicians, provided orchestration.

The program consisted of two acts, with songs performed with such titles as "On With The Show," "What A Life," "My First Love" and "Pass The Buck." Dancing soldiers, "The 317th Serenaders" and an accordion specialty by Pvt. Tomaszowski were on the bill.

"Yes, there was some entertainment on the post," Whiteside recalled, "but my relaxation was going to the PX after the dinner hour and just hanging out with friends having a few beers. We went to a number of movies, I think there were two theaters on the post." Whiteside remembers, "I probably went into town (Paris) about three times a month. We only received fifty dollars a month, so the money ran out quickly. The USO was on the town square on the second floor. I remember going there a few times, but they were very fussy with the girls there, could not leave the room with a soldier for any reason...We did picnics, car rides, carnivals and town stuff." [264]

---

[263] Howard Koenen, personal interview, October 30, 2009.
[264] Ibid.

There were plenty of opportunities to meet local girls both on and off the base. Dances were held regularly at the camp and at the USOs in both Paris and McKenzie and there were several married ladies in both towns who were more than willing to organize afternoon teas and other chaperoned get-togethers so that the local single ladies could meet — and hopefully date — the glamorous soldiers.

War may be hell, but it also provides a good opportunity for romance. Local women in Paris and McKenzie still remember those days with wistful fondness, for it was an exciting time for meeting and dating men from all corners of the globe.

"Some people in town didn't like the soldiers because they were Yankees and they all talked so funny," Jeanne Townsend remembered. "But we liked them and they liked us. A lot of the soldiers liked the way we Southern girls talked. I remember they liked the way we said 'chocolate.' For some reason, they thought that was cute." [265]

With her position as secretary to the camp officers, Townsend was in a perfect position to meet and greet soldiers. "There were a lot of enlisted men in our office. I remember meeting some from Switzerland, Lebanon, Poland. And that was an education in itself. To me, it was interesting and I learned a lot." [266]

Townsend was raised by her grandparents, who "really didn't want soldiers in this town and who didn't want us to date soldiers." But she did eventually date a soldier, who was "Polish and Catholic. It was pretty hard for me to introduce him to my grandparents, I must say." Townsend said she enjoyed meeting the boys from all over. "I got to where I could tell if they were from New York or New Jersey," she said, "just by the way they talked." [267]

Before McKenzie had an organized USO, 'home parties' were organized as a way for soldiers and young ladies to meet. Kathleen King, wife of then Mayor Glen King, was especially good at playing cupid. Living in a large, elegant home on Magnolia Street, the Kings had the perfect spot to play host. "The house was very grand in those days and Mrs. King delighted in entertaining with parties...She was one of the

---

[265] Jeanne Townsend, personal interview with author, April 2010.
[266] Ibid.
[267] Ibid.

first to organize social gatherings and dances for the young soldiers..."[268]

Maxine Bersey was a primary beneficiary of Mrs. King's matchmaking, as the mayor's wife was responsible for her meeting her future husband, Kenneth Bersey. "Aunt Kass, as I called her, was so persistent. I was so shy and she just kept calling me to come and meet him," she said. "There was a whole group of soldiers who were looking for girls," Bersey said. "And they liked the Southern girls, too. That was the first thing they were looking for."[269]

For local girls, it was a time wrought with trepidation for the drama and heartache of war, but exciting, as well. "Paris was alive with energy," Jerry Ridgeway of Paris said. "Teenage girls wanted to look older to attract the soldiers and I must say the soldiers were never out of line." Ridgeway recalled there was a neighborhood skating rink near Wood Street in Paris "where the music was happy and the soldiers loved to come. As teens, my best friend and I had learned to skate dance really well. We skated with those young soldiers every chance we could. They were so exciting to all of us teen girls."[270]

THE GREYSTONE AND A ROTHSCHILD

Nell McBride of Paris was in the perfect spot for meeting soldiers, working as a waitress at The Greystone, the fanciest hotel in Paris and for many miles around. The Greystone boasted all the amenities of your big city hotels. White-gloved porters helped guests with their luggage, waitresses in uniforms served them in the elegant dining room, at which chefs provided the finest dining. A florist shop, lounge with grand piano, and comfortable rooms in the three-story hotel rounded out The Greystone experience.

"Were there ever a lot of soldiers there! Oh me! It was quite a lively place," McBride said. "Some of the engineers who helped build the camp stayed there and then after the camp was built, soldiers and officers began staying there, too. But a lot of soldiers and officers would just come there to eat or to sit in the lounge and listen to music, visit." McBride also remembers going to the camp to dances "which was sort

---

[268] *The McKenzie Banner*, March 19, 1997.

[269] Maxine Bersey, personal interview with author, January 27, 2010.

[270] Ibid.

of glamorous. They had big bands out there and you wore an evening gown." [271]

A soldier from a prominent New York family was stationed at Camp Tyson — Howard Rothschild — and his parents came to visit him frequently. And, of course, The Greystone was the only place locally that such a wealthy couple would find adequate for their stays.

"A young Army private, one of the Rothschild family, was there a lot," McBride recalled. "He was such an odd fellow. He would only let Mildred McLain wait on him all the time. His parents came quite a lot to see about their little boy." [272]

The late Harry Neal, a Paris musical performer who later wrote a self-published book, *Wave As You Pass*, remembered Rothschild in his book.

"Into all the confusion of my musical ambitions came the first professional artist I ever knew. Howard Rothschild was a sensitive and sophisticated artist." Neal, who was a young teen during the war years, remembers that "Howard came to Miss May, my teacher, to ask if he might use her piano for practice on his days off. She graciously opened her home to him and also asked him to hear me play, thereby giving my life a turning point for which I shall always be grateful." [273]

Mike Freeland, who now lives in Hopkinsville, Kentucky, worked at The Greystone as a teenager. Freeland's family lived in Buchanan near Kentucky Lake (a few miles from Paris) and he stayed in a less expensive Paris hotel across the street — The Caldwell Hotel — while he worked at The Greystone.

"It was very fancy — we wore white jackets and the whole bit," Freeland said. "The Greystone was very busy and pretty special. We had two excellent chefs working there and it was very expensive. The officers used to eat there and they were big tippers, I remember. They were the only ones who could afford to eat there." [274]

Soldiers did come to The Greystone to relax in the lounge "and compete for dates," Freeland said. "It was a grim time for the country, but it also was a good time for dating." Freeland later wrote a book, *Blood River to Berlin* in which he talks of his war experiences. [275]

---

[271] Neil McBride, personal interview with author, October 6, 2009.

[272] Ibid.

[273] Harry Neal, *Wave As You Pass* (J.B. Lippincott Company, Philadelphia and New York, 1958), page 42.

[274] Mike Freeland, telephone interview with author, April 7, 2009.

[275] Ibid.

Paris during the war "was an absolutely busy time. Sometimes you couldn't walk on the sidewalks for all the soldiers. And it wasn't just military. There were many more people in town then." The Greystone had a jukebox, he said. "You'd put in a nickel and out would some 'Stardust' and Glenn Miller. You can't talk about that period without talking about the music. You couldn't really have a war without music." [276]

SHADY PARK

The former Shady Park Trailer Camp on E. Wood Street, Paris, was a popular spot for locals and soldiers alike — so popular that even Country Star Roy Acuff used to go there after the war.

The trailer camp was owned by Bob King, who had been an MP in the U.S. Army during World War I. "My father went to Camp Tyson, explained we had a family business, and that soldiers would be welcome, and be given fair treatment," his son, Wayne King recalled. [277]

The big night at Shady Park was Saturday, he said. "On Saturday nights, we had dozens, maybe hundreds, of soldiers visit Shady Park." King, who now lives in Lakeland, Tennessee, recalled that the camp — which was located on 4 ½ acres of land, included small houses to rent, a restaurant, dance hall and living quarters for their family.

A teenager then, King was impressed with the soldiers who frequented the park. "We charged five cents for soft drinks; ten cents for regular beer; 15 cents for premium beer. A large platter of country ham, with extras added, was 25 cents. We had a Wurlitzer which played music, five cents per record. And there was no cover charge for using the dance hall." [278]

To ensure control on Saturday nights, King recalled, the Camp Tyson Military Police would send a 'paddy wagon' to the front door of Shady Park around 8 pm. "They would open the doors of the paddy wagon, check outside, go inside the building and fill up the paddy wagon with soldiers who they thought needed to go back to the camp and take them back to Camp Tyson. About an hour later, the procedure was repeated, until the business was closed for the night." But the

---

[276] Ibid.
[277] Wayne King, letter to author, May 1, 2009.
[278] Ibid.

AS IF THEY WERE OURS

'paddy wagons' were largely unneeded, King said. "We NEVER had a fight from the soldiers or problems with them." [279]

MOONSHINE AND GOOD OLE' BEER

Among those who were happiest that Camp Tyson located in Henry County, according to the late William Crosser of Paris, were "the bootleggers. The bootleggers were real happy because there was a long list of bootleg joints around here and the soldiers figured out where they were." [280]

Nancy Cate, 80, of Paris lived within throwing distance from the camp and can recall seeing "soldiers crawling through the big culverts to go to the beer joint on the other side of us." The huge culverts are still under the roadway on Steele Rd. and soldiers would crawl through them after hours, she said. [281]

"There was a beer joint in Routon that was called 'The Trees' and that's where they would go at night," Cate said. "Of course, they weren't supposed to — it was way after dark. Even when we couldn't see them, we could hear them moving around in the woods on the way to the beer joint." [282]

Maj. William Jones of Tavares, Fla., was stationed at Camp Tyson with the 308th Coast Artillery Barrage Balloon Battalion in 1942. One of his most vivid memories of the place involved a walk one night into Paris.

"There was a hill, a wooded hill, that started up right at the edge of the road, a dirt wagon road went up the hill. As we passed the dirt road an old man was just coming down," Jones said. "Showing our ignorance and being smart alecks we jokingly asked if he had any 'white lightning.' [283]

To their surprise, the man said yes. "We followed him to his small, old house, around back where he picked up a shovel and we followed him 20 or 30 feet on up the hill when, of a sudden, he stopped, scraped away some leaves dug just a few inches down and uncorked a gallon jug of uncolored liquid." [284]

---

[279] Ibid.
[280] William Crosser, telephone interview with author, August 6, 2009.
[281] Nancy Cate, personal interview with author, May 28, 2009.
[282] Ibid.
[283] Major William Jones, letter to author, June 25, 2009.
[284] Ibid.

Jones said, "As I recall, we paid him $2 for the jug. Neither one of us were drinkers, not even beer. We both were just 19-year-old farm boys. I from New Jersey and I believe he was from Pennsylvania. We started nipping the jug and I do remember that occasion. Neither of us remembered being in Paris that night or how we got back to camp." [285]

---

[285] Ibid.

Chapter 12

The Day Jeanette MacDonald
Came To Town

OF all the stars who entertained the soldiers at Camp Tyson, one of the brightest was Jeanette MacDonald.

As a star of stage and the silver screen, Miss MacDonald's appearance was covered extensively by the local press. Crowds of starry-eyed locals — eager for a glimpse of Hollywood glamour — awaited her arrival and departure on the Pan-American at the huge L&N train depot, even as heavy rain fell down on Paris.

Jeanette MacDonald's husband, Lieutenant Gene Raymond, was in active service in Europe, and the star volunteered to entertain the troops as a way of honoring her husband.

The humidity of the August 1, 1942, morning gave way to a downpour later that night. It rained so hard that the entertainers had to seek shelter, as the show was moved from the camp's outdoor Motor Park to the indoor Post Theater. [286]

The show was again interrupted when the thunderstorm "played havoc with the current," but, trooper that she was, MacDonald "sang through the intense blackness," noting that "Our boys fight through the dark, and certainly we can sing in the dark." MacDonald said that nothing could keep her spirits down, as she had received a letter from her husband while in Paris. In his letter, her husband told her, "You keep 'em singing while I keep 'em flying." [287]

The star arrived in Paris at 10:48 a.m. from Camp Robinson in Little Rock, and since she would not be departing until the next day, immediate attention was paid to finding a suitable spot for such a major star to spend the night. [288]

A large downtown hotel, The Laureighn, was selected by the camp's officers.

---

[286] "Movie Star Charms Tyson Audience," *The Paris Post-Intelligencer*, August 2, 1942, front page.

[287] Ibid.

[288] *The Paris Post-Intelligencer*, August 2, 1942.

The late Laureighn Coplen, for whom the hotel was named, told *The Paris Post-Intelligencer* in 1967, that the camp's commanding officer, General Maynard, called her to personally request that MacDonald stay at the hotel.

"Well, I told him we just couldn't," she said. "She had all these companions and sound men and all that. Why, she wanted a suite of rooms." But General Maynard insisted and Mrs. Coplen "scurried about, trying to find only the best for her special guest. She even took furniture out of her own home to put into the group of rooms which comprised Miss MacDonald's suite." [289]

To accommodate the star's local sojourn, "We just let the rest of the hotel go," Mrs. Coplen said, noting that her staff was transformed into personal secretaries for MacDonald, screening phone calls and catering to her needs. "You wouldn't believe how that switchboard was just ringing off the wall," Mrs. Coplen said. "We had to tell them that she was not available." [290]

During her visit, MacDonald visited several Paris stores, lunched and had dinner at the Plaza Café, and enjoyed the bouquets of flowers that were placed in her rooms from local admirers. People who saw her during her visit recall her beauty — and the golden glow of her hair. "Oh, she was a gorgeous creature!," Mrs. Coplen recalled during the 1967 interview. [291]

And even 67 years later — in 2009 — a couple of Parisians could add their memories. "She had a certain magnetism," Marilyn McFadden of Paris remembered. "But what I remember most was her hair. I don't remember if it was red, auburn, or gold, but you felt like you needed sunglasses because it was so bright." [292]

McFadden, who was a young teen at the time, said her family was among the crowd that greeted MacDonald when she arrived at the depot. "You can imagine how we felt," she said. "And as I recall there was a fair-sized crowd that was there. When she turned to smile at us — well, she had a lot of personality." [293]

---

[289] Laura Coleman, "No More Memories To Be Made At Laureighn Hotel," *The Paris Post-Intelligencer*, undated month and date, 1967, property of Stephanie Routon Tayloe.

[290] Ibid.

[291] Ibid.

[292] Marilyn McFadden, telephone interview with author, August 28, 2009.

[293] Ibid.

Franklyn Thompson of Paris said his family went to see MacDonald as she left The Laureign Hotel. "I was just a little boy, but I remember people saying there was a movie star staying there. If I remember correctly, it was raining, but we stayed out there anyway. She was on her way to the camp and we caught a glimpse of her as she came out the door."[294]

The local newspapers raved about MacDonald's performance — and her appearance. "The singer was attractive in a formal of pale green mouseline de soie that enhanced the charm of her red-gold hair."[295]

"The entertainer gained the affection of her audience from the very beginning of the program," *The Parisian* reported, by accepting song requests and lending her soprano to such favorites as "Ave Maria" and her song for which she was most famous, "Indian Love Call." Recognizing the enthusiasm of the overflow crowd of officers, enlisted men, their wives and civilian guests, MacDonald insisted on presenting a second concert. And earlier that evening, she had sung a few numbers for patients at the post hospital.[296]

But it was what MacDonald did earlier that afternoon that charmed the people of Paris most.

At an earlier visit to an army camp in California, MacDonald had told the crowd she was going to be entertaining at Camp Tyson, near Paris. A soldier there asked her to deliver a message to his sweetheart, who lived in Paris.[297]

"It was agreed that the Hollywood Victory singer would call that friend over the telephone on reaching this city, but Miss MacDonald decided that a personal visit would be much more satisfactory," according to *The Parisian's* account.[298]

McFadden recalled that MacDonald had inquired and found out the soldier's sweetheart, Ruth Humphreys, worked as the office assistant to her brother, Dr. K.B. Humphreys, Sr. "He was a dentist and he had an office in the Commercial Bank building," McFadden said. "I

---

[294] Franklyn Thompson, telephone interview with author, September 15, 2009.

[295] "Jeanette MacDonald Gets Third Letter In Three Months From Airman Hubby While in Paris," *The Parisian*, August 7, 1942.

[296] Ibid.

[297] McFadden interview.

[298] *The Parisian*, August 7, 1942.

remember that Jeanette MacDonald went up to his office to visit her while she was here." [299]

The Parisian noted that Ruth Humphreys had no idea that she would be receiving a visit from the movie star that day. "She went forward...with her best professional air, to greet a would-be customer, only to find that she was receiving a personal call from the lovely Jeanette." [300]

Needless to say, *The Parisian* reported, Miss Humphreys was "thrilled." [301]

[299] McFadden interview.
[300] *The Parisian*, August 7, 1942.
[301] Ibid.

Chapter 13

Death In An Army Camp

WE may never know what actually happened in the darkness of the evening on July 31, 1943, but we do know the tragic end result.

A black Camp Tyson soldier was shot and killed by white soldiers in a tourist camp just yards away from the camp and for decades has lain in an unmarked grave in a Paris cemetery.

In the aftermath of the shooting, Pvt. Herman Hankins of Danville, Va., was accused of peeping through the bedroom window of a white soldier's wife, but at least one former soldier who knew him doubts that occurred.

"All I know is that Herman was murdered and that it was under mysterious circumstances," according to Wilson Caldwell Monk, who was a Master Sergeant in the 320th Anti-Aircraft Barrage Balloon Battalion at Camp Tyson. [302]

Monk, who now lives in New Jersey, visited the grave in Maplewood Cemetery after Hankins' funeral and said the shooting and its aftermath was a time of great trepidation for black soldiers at Camp Tyson. He said he couldn't remember now whether the black soldiers were allowed to attend the funeral, but does remember visiting the grave with several of his buddies. Monk still has a photograph of the grave, with a number of flowers laid across it.

"I didn't believe what Herman was accused of," Monk said. "He wasn't that kind of guy. He was a very respectable young man and would never do anything like that." [303]

Hankins was a member of Battery B of the 320th Battalion.

The shooting death was made public to the local media and both Paris newspapers — *The Parisian* and *The Post-Intelligencer* — reported the incident. The newspapers apparently published prepared news releases from Camp Tyson, since both of their articles were identically worded.

---

[302] Wilson Monk Caldwell, telephone interview with the author, April 13, 2010.
[303] Ibid.

"Failure to halt when commanded cost Pvt. Herman Hankins, 25, Camp Tyson soldier, his life about 10:00 o'clock Thursday night," both articles begin. [304]

According to the articles, Hankins was "surprised prowling" near a home in Routon, the small town which adjoined Camp Tyson. As related by the newspapers, Hankins was prowling near the home of William A. Barringtyne and that Mrs. Barringtyne's husband and two friends "had been watching the place since she had reported a man peeping in her window about 9:30 that evening." [305]

When Hankins "fled on being twice commanded to halt, one of the men fired at him with a single shot .22-caliber rifle," according to the articles, which went on to state that the incident was reported to the Military Police stationed at Gate 1. The ambulance was called, but Hankins "died from hemorrhage before he reached the Station Hospital." [306]

What the articles didn't reveal, however, was that the ambush was perpetuated by white soldiers: members of the camp's military police unit, in fact.

FULL STORY REVEALED

A document in the National Archives and Records Administration (NARA) in Washington, D.C., reveals the full story, as related from an initial investigation by Army officers.

According to the "Report Of Death Of Pvt. Herman Hankins," filed July 31, 1943, Hankins was shot by Pvt. Murphy Price, Jr., a member of the 316th Anti-Aircraft Barrage Balloon regiment.

"Investigation revealed that about 15 families of Camp Tyson soldiers live in a group of cottages near the highway at Routon" [307], a half-mile from the camp. The cottages were part of a tourist area owned by Pearl Routon of Paris. "All the families are white families," according to the report. "During the past four months, the Provost Marshal has received numbers of complaints from these families about prowlers in

---

[304] *The Parisian*, August 2, 1943; *The Post-Intelligencer*, July 30, 1943.
[305] Ibid.
[306] Ibid.
[307] "Report Of Death Of Pvt. Herman Hankins," July 31, 1943, NARA, RG 319, Entry 47, Bx 1319, Ordinance plants, Camp Tyson (folder 004.4) Army Intel Decimal File (1941-1945).

the vicinity, and has made numerous attempts to apprehend the culprits, without success." [308]

The report states that the wife of Sgt. Albert Lovin, a member of the Military Police detachment, said she had observed three Negro soldiers pass along the highway in front of the house about dark "and heard one of them state that 'we will be back.' [309]

After dark, the wife of Pvt. William Barrentone — spelled differently in the document than it is in the newspaper articles — another member of the 316th, "saw a negro soldier peeping in the northeast corner of her room. She reported the incident to her husband, as well as to Pvt. Murphy Price and Pvt. Roy L. Smith, "all of whom immediately took up guard positions outside the house." [310]

At about 9:40 p.m., the three soldiers "observed a prowler approaching the house on his hands and knees, crouching near the ground each time a car would approach in his direction. When the prowler was about 25 feet from the house, the men flashed a light on him and attempted to arrest him," according to the report. "The soldier ran and Pvt. Price, who was armed with a single shot .22-caliber rifle, called 'Halt' twice. When the prowler failed to halt, Pvt. Price shot him one time." [311]

The "prowler" continued running, crossed the main highway and fell to the ground. Pvt. Barrentone drove to the nearby Gate 1 and reported the shooting to the MP on duty at the guardhouse. An ambulance was called to where the prowler had fallen and it was found that he still had a pulse at that time. "The ambulance arrived a few minutes later, but Subject was dead when he arrived at the Station Hospital." [312]

INVESTIGATION LAUNCHED

Within moments, the Officer of the Day, the Field Officer of the Day and the Provost Marshal arrived on the scene and launched the investigation.

According to the report, Pvt. Hankins' next of kin was notified and a Board of Officers was appointed to fully investigate.

---

[308] Ibid.
[309] Ibid.
[310] Ibid.
[311] Ibid.
[312] Ibid.

Officers at Camp Tyson were fully cognizant of the furor that could be caused with such a racially-charged event. "No repercussions from the incident, either pro or con, among the negro personnel at this station" were reported, according to the report, "although a very careful watch has been maintained." [313]

UNMARKED GRAVE

The incident might have been lost to history were it not for two seemingly unrelated occurrences. Bill Davison, son of former Camp Tyson soldier George A. Davison, submitted an undated article relating the death of Hankins for this book. [314]

Around the same time, former Paris Parks Employee Don Williams found a log of all burials performed at Maplewood Cemetery in Paris from 1927 through 1952. In the book, he found a notation that two burial plots had been purchased by Camp Tyson.

According to the notation, the burial plots were purchased for soldiers in the 320th Barrage Balloon Battalion.

The burial log listed no names of the deceased in the plots. That is unusual, but does not necessary mean there were no bodies buried there.

Using a 'witching' instrument, Williams determined there was a casket in one of the plots, located in a segregated section of Maplewood in which black people had been buried.

Could it be possible that the article kept in Davidson's scrapbook was related to the unmarked grave in Maplewood?

The genealogy room at the Paris-Henry Co. Library maintains copious lists of all burials that have occurred in the county. A search in the book "Henry County, Tennessee, Black Funeral Home Records, 1912-2002" found a notation of burial for Herman Hankins in Maplewood Cemetery. [315]

Even with all the information we have, however, mysteries still remain. Why was Pvt. Hankins so hastily buried in Maplewood? Why was his body not returned to his home in Danville, Va.? Why was he buried in an unmarked grave?

---

[313] Ibid.

[314] "Tyson Soldier is Fatally Wounded," undated article from unknown newspaper, property of Bill Davison.

[315] *Henry County, Tennessee, Black Funeral Home Records, 1912-2002* (Henry County Genealogy Society), page 52.

AS IF THEY WERE OURS

It is the unanswered questions that make the incident all the more tragic.

Chapter 14

"The Deuces"

The African-American Experience

IN many ways, the Camp Tyson experience in the African-American community of Henry County was the same as in the white community. As did the white community, black households opened their doors to the soldiers and their families, many long-standing marriages resulted from the Camp Tyson experience, lasting friendships developed and people were just as excited to see the well-tailored soldiers — in this case, black soldiers — as they walked down the streets.

But in more important, intrinsic ways, the experience was a world apart from that of the white community. For many, the Camp Tyson experience magnified the separate nature of the segregated South. This was before President Harry Truman integrated the U.S. Army; black soldiers were housed and trained separately from whites at the camp, within their own segregated units.

Beyond that, the black soldiers were expected to visit only the black sections of Paris when they came to town, had their own USO, and their own social lives which were separate and apart from the white community. And their families stayed in separate housing when they visited the camp. For many black soldiers — especially those from Northern urban areas — this was a difficult arrangement and it did lead to some racial tensions at the camp and in Paris.

According to Linda Crutchfield and Dorothy Vaughn, sisters who were brought up in Paris, the black soldiers had the nickname of "The Deuces" and made quite an impression on their community — especially the women.

Crutchfield, of Paris, said, "I know the black people were excited to see the black soldiers in town. They'd come to town every weekend, but when the women would pass them on the street, they would stop what they were doing and let the women pass. They had real respect for the women and children. They made quite an impression on everyone." [316]

SWEETHEART OF THE BATTALION

Vaughn, who now lives in Chicago, had the enviable experience of being named "The Sweetheart of the Battalion" at the black USO when she was 10 years old. "I remember the dress I wore and the candy and attention I got," she laughed. [317]

Before she retired, Vaughn was director of alumni relations at Hunter College near Chicago, and remembers that evening and the USO well. "I wore a white satin evening gown with a red heart in front. I'm sure someone made it for me. I remember that dress as well as what I wore yesterday." [318]

The USO building was in a large still-existing structure on Rison Street in Paris. "I can remember how big it was inside there. Coat checks on each side of the entryway as you walked in," Vaughn said. "My mother was active in the USO and several women from our community would volunteer as hostesses and make sure the food was wholesome for the soldiers. The black soldiers "were in the city all the time. It was part of our lifestyle," she said. "And they made quite an impression. Men in uniform will attract women in a small town. And black people didn't see that many black men in uniform in those days." [319]

Vaughn left Paris when she was 16 to attend college and said there were racial tensions when the camp was open. Many of the black soldiers were from the North "and had never been exposed to overt racism, so that was difficult for them, but they stuck it out." Still and all, she said, "Paris is my home. I had a great life, despite segregation." She credits her parents for much of that. "They taught us manners, how to survive in the world and not let it get you down, even during that time." [320]

Lester Teague owned the USO and his son, Lester, Jr. still owns the building, for many years was rented out for dances and other events.

---

[316] Linda Crutchfield, telephone interview with author, June 8, 2009.
[317] Dorothy Vaughn, telephone interview with author, June 9, 2009
[318] Ibid.
[319] Ibid.
[320] Ibid.

"I was small when the USO was there," Teague said, "but I can remember the dances in there and the jukebox. I also remember the soldiers used to give us kids money, I mean coins, and that made us pretty happy."

Camp Tyson was an important part of the Teague family's life, he said. His father helped build the camp and was one of the men who helped tear it down after the war. Lester, Jr. worked for a few years at the Spinks Co. that purchased the camp when the war was over. [321]

The late Mary Will Gardner of Paris, was one of the local ladies who owed her lengthy and happy marriage to Camp Tyson. Gardner was a teacher at the Henry County Training School, but also worked at the USO as a hostess during the war.

At the age of 100, Gardner said the wartime was "pretty exciting. You would see the soldiers in town all the time, especially on Sundays. They used to come to the church services and people would invite them to their houses for Sunday dinner." [322]

She was working at the USO when she first saw her late husband, Oscar. "They were having a dance over there and he came up and started talking to me. He asked for my phone number" and the two began dating shortly after that. [323]

Earlie Pierce of Paris lived on Rison Street and remembers the excitement the soldiers caused and that it almost led him to a life of crime. "I was 8 or 9 years old and we lived close to the railroad. We liked to watch the troop trains go by. We also were close to the USO and a friend and I decided we'd shine shoes for the soldiers, but we didn't have any money for polish." [324]

Pierce laughed at the memory. "We decided we'd go to Woolworth's and steal some polish. I got two or three bottles and we got outside, but my buddy decided we needed one more, so he went back in and got caught. The owner said you bring your buddy back in and I'll get y'all in jail. I didn't want to go to jail, so I ran off. I never did shine any shoes for any soldiers." [325]

---

[321] Lester Teague, Jr. telephone interview with author, July 16, 2009.
[322] Mary Will Gardner, personal interview with author, May 4, 2004.
[323] Ibid.
[324] Earlie Pierce, telephone interview with author, July 17, 2009.
[325] Ibid.

Aaron Dobbs, Jr. of Paris remembers his father and mother speaking of Camp Tyson frequently over the years. His parents, Aaron Dobbs, Sr. and Joy Randle, met when the camp was open. His father was a Staff Sergeant in the all-black 319th Headquarters Battery at Camp Tyson, stationed at one of the motor pools there. "Dad always said that certain jobs were relegated to the black troops," Dobbs said. "Cooks, mechanics, mess hall, clean-up crews. Those were the kinds of jobs they were assigned out there."[326]

Dobbs, Jr. made his career in the Army, working in military intelligence, and understands how things were done before the Army was integrated. "Things could get pretty rough," he said, "but when the camp was open, things in Paris were pretty exciting."[327]

As in the white community, the Camp produced plenty of side businesses for the black community and a boost in the economy. "A lot of businesses grew up because of all the soldiers around. You could get a job cleaning cars for the soldiers. Women in the community would clean uniforms, do their laundry. Also, you had a lot of liquor, moon-shining businesses and gambling parlors that opened up."[328]

Areas in Paris that were especially lively were the "black bottom" of E. Washington and W. Blythe Street, both of which were home to several beer joints and restaurants. The Brown Derby on Washington St. was a popular hangout and was where Dobbs' parents met. The Tip Top Café at 418 W. Blythe St. was another. Both of those establishments are now closed.

Paris "was like a magnet" for the soldiers, he said, "especially on pay days."[329]

The Tip Top Café was owned by John Wesley Bobo, a cousin of Roland Atkinson. "It was a hopping place and I worked there when I was a teenage boy, waiting on tables,"[330] according to Atkinson.

The Tip Top was a fancy establishment, with waitresses dressed in white informs, aprons and small white hats. "It wasn't a juke joint. It was a nice place and it was packed with soldiers. And the soldiers were respectful because John would tell them to be that way. They called

---

[326] Aaron Dobbs, telephone interview with author, February 24, 2010.
[327] Ibid.
[328] Ibid.
[329] Ibid.
[330] Roland Atkinson personal interview with author, February 28, 2010.

him 'Papa John' and he didn't play. And he was packing — he had a piece on him all the time." The soldiers liked it at the Tip Top, Atkinson said, "because there was good food and they felt safe there." [331]

When Atkinson was in 8th grade, he remembers his classes at the Henry Co. Training School going to Camp Tyson to sing for the soldiers. "We had concerts out there. Good music. Patriotic music. There were predominantly black soldiers in the audience when we sang out there." [332]

Both Dobbs, Jr. and Atkinson say that the Camp closed because of the changes that Camp Tyson brought about in the black community.

"One of the reasons the Camp closed," Dobbs said, "is because the white upper crust was peeved that they couldn't get black women to be their maids anymore. Their maids had married soldiers and had money of their own; they didn't need the work anymore." [333]

Atkinson agreed, noting "The powers that be didn't like the fact that black women weren't willing to be maids anymore." [334]

STAYING AWAY FROM PARIS

But there were racial tensions in Paris. Both William Dabney of Roanoke, Va., and Wilson Caldwell Monk of Atlantic City, New Jersey — both of whom were members of the 320th Anti-Aircraft Barrage Balloon Battalion — said they did not come to Paris while they were stationed at Camp Tyson.

"We were told to stay away from Paris; that it was a terrible place for blacks," Monk said. [335]

Monk was a friend of Pvt. Herman Hankins, who was killed in Routon by a white soldier, and he said that event placed a pall over the black soldiers. "We never went to Paris. It was disheartening. You wear the same uniform, but you're not accepted by whites." When he and his buddies socialized, he said, "We went to Memphis or Jackson or watched movies, played cards on the post." [336]

Dabney was a Southerner and was used to segregation. "I was used to blacks being here and whites being there. The Northerners looked

---

[331] Ibid.
[332] Ibid.
[333] Aaron Dobbs interview.
[334] Atkinson interview.
[335] Wilson Caldwell Monk, telephone interview with author, April 13, 2010.
[336] Ibid.

at it differently." He and his friends did not socialize in Paris, he said. "We went to Paducah, Kentucky, on weekends. We wanted to go to a bigger city, where there were more African-Americans to associate with." [337]

A DOWNTOWN INCIDENT

There are a few members of the black community who remember an incident in downtown Paris that made people nervous.

Dobbs said he remembers his mother telling him that some black soldiers, with their white officer, had come downtown. The white officer started walking into one of the downtown soda fountains and the black soldiers stopped at the door. The officer told them to come on in and they said they weren't allowed. The white officer then went in and asked the proprietor of the soda fountain if his men could come in and was told no. The white officer raised a fuss and the proprietor then said they could order sodas but had to drink them outside on the sidewalk.

"My mother said things really got tense around town after that. People were afraid that some kind of disturbance would materialize, but it never did," Dobbs said. [338]

Maynard Cook of Paris also remembers the incident being discussed. "I believe it was Sullivan's Drug Store where that happened. There was a disturbance about the policies of the town and everybody got riled up for a time." [339]

Another person who remembers the incident is Henrietta Holmes of Paris. "The white commanding officer told the owner to make sure his men were served and that was a big deal. Everyone in Paris was upset and scared that something was going to happen, but it never did." [340]

Cook remembers being fascinated with the soldiers as a little boy. "I remember Lucien Freeman had a barber shop on Brewer Street, upstairs on the second floor. We'd come into town from out in the country and when I was at the barber shop, I used to like to look out the window and stare at all the soldiers on the streets below." [341]

---

[337] William Dabney, telephone interview with author, June 8, 2009.
[338] Aaron Dobbs interview.
[339] Maynard Cook, telephone interview with author, March 30, 2010.
[340] Henrietta Holmes, personal interview with author, April 14, 2009.
[341] Ibid.

AS IF THEY WERE OURS

The late Clarence Clark of Paris said Camp Tyson opened up jobs for black people, but only menial ones. "I worked at Camp Tyson at the officers' club on Saturday nights. This was before integration, of course, so it was the white officer's club. I waited on tables, served food. All the people who worked there were black. That's how it was then." [342]

MUSIC WAS INTEGRATED

Although most areas of the camp and the local communities were segregated, musicians at Camp Tyson did find a way to share their love of music.

The late Tom Lonardo, director of the Camp Tyson band, recalled although there were no blacks in the official camp band, black soldiers had their own band.

"I remember that the black band members would come to our barracks at night and we had jam sessions," Lonardo said. "We had more fun doing that." [343]

The late Lela Capps also worked at the Camp, as a cook. "I did enjoy working for the soldiers. It was a busy time. We cooked good meals, vegetables, meat, desserts." [344]

On the other side of the coin, however, Nona Moore said her father's experiences at Camp Tyson were not bad. Her father, W.T. Millikan, was a soldier at Camp Tyson and is buried in Greenwood Cemetery in Paris. In an interview with the author in May of 2009, Moore said she remembers him talking about going to the USO on Rison Street and having good times. "He made a lot of friends at the camp and seemed to have pretty good experiences," she said.

SALUTING WOMEN SOLDIERS

R.L. "Shorty" Hutcherson, who grew up near the small town of Cottage Grove several miles from the camp, said he remembers the crowds of soldiers when his family would come into Paris to shop. "It

---

[342] Clarence Clark, telephone interview, February 15, 2010.
[343] Tom Lonardo to Susan Gordon for Tennessee History Society "Home Front Project" April 17, 1992.
[344] Lela Capps interview with Susan Gordon, April 17, 1992.

was something else. There were so many soldiers, you couldn't hardly walk down the street." [345]

The black community was very excited about seeing the black soldiers, he said. "It was something I really enjoyed. I remember it was something else. You used to see a lot of women soldiers, the WACS, walking through town. They were all white, but all the men had to salute them, black and white. That was the prettiest sight I ever saw." [346]

For Valeska Surrette Hinton, Camp Tyson was a stepping stone to an exalted career.

Hinton was born in 1918 in Paris, and while the camp was open, she counseled and tutored black soldiers there. She also helped found the black USO.

After the war, she married Robert Hinton in Peoria, Ill., and began working for the USO there and for a community center. She became active in community affairs and was nicknamed "The Mother of Peoria's Civil Rights Movement." She was the executive director of the city's Human Relations Commission and then moved to Washington D.C. to work for the U.S. Commission on Civil Rights until 1983. [347]

The Valeska Hinton Early Childhood Education Center in Peoria is named for her. She died on Sept. 22, 1991.

---

[345] Shorty Hutcherson, personal interview with author, April 8, 2010.
[346] Ibid.
[347] www.valeskahinton.psd150.org.

Chapter 15

"We Were There. We Did Our Part"

O NE of the saddest aspects of the Camp Tyson story — if not the saddest — is that the members of the 320th Anti-Aircraft Barrage Balloon Battalion did not receive the recognition they deserved for their bravery and fearless hard work.

The 320th was the only all-black military unit to storm the beaches at Normandy on D-Day and they were trained at Camp Tyson.

But their story had been lost to history until recently. Even those who should have known — the people who lived in Henry County, both white and black — were completely unaware of the storied legacy of the 320th.

Like all black military units of that era, members of the 320th had more than the physically tough training and military regimentation to face. At the same time they combated those rigors, they also had to confront the ugliness of segregation.

The U.S. Army would not be integrated until 1948 by President Harry Truman, and for those soldiers who were stationed in Southern states, the tribulations of segregation were compounded. For them, there was no respite from the separate existence in the army camps since the nearby towns also were divided along color lines.

"At that time, we were like two armies — black and white," Wilson Caldwell Monk, 90, recalled. Monk was a Master Sergeant for the 320th and was trained at Camp Tyson. "Generally, black troops occupied the rear part of the camp and had a separate existence." [348]

He is one of the very small handful of still-surviving members of the 320th who can be located today.

Other members of that small club are William Dabney, 80, of Roanoke, Virginia; and Henry Parham of Pittsburgh.

Of those three, only Dabney has thus far been publicly honored for his achievements and the award he received was a big one: Dabney received the Legion of Honor, presented to him by President Barack Obama, in a special ceremony in Paris, France, on June 6, 2009.

---

[348] Wilson Caldwell Monk, telephone interview with author, April 13, 2010.

At the time of the presentation, it was believed that Dabney was the only surviving member of the 320th. With the help of Alice Martine-Mills, 60, a Frenchwoman who teaches English in Paris, the White House had arranged for Dabney's inclusion rather hastily after learning of him.

After seeing an old photograph of members of the 320th underneath a huge barrage balloon, Mills took it upon herself to begin researching the battalion. She traveled across France, collecting stories of her fellow countrymen who remembered the black soldiers and she did research in the Library of Congress. She said it was important to her that the soldiers get the recognition they deserved. "All my life I've felt that because I'm not speaking German, because I'm free, it's because of the Americans," she said. "So I had to get the stories of the black soldiers."[349]

The lack of recognition is one frustration upon many the veterans had lived with all their lives. Monk said, "At Camp Tyson, we had separate everything. Separate service clubs, separate barracks. We didn't mix with the whites. I had read and heard how the South was toward Negroes, but it was disheartening to experience. You know, you're wearing the same uniform, but you're not accepted by the white soldiers."[350]

NO BLACKS IN "SAVING PRIVATE RYAN"

In the History Channel documentary, "A Distant Shore: African-Americans at D-Day" several black veterans, including Dabney, shared the same thoughts.

It was especially galling for Dabney to watch "Saving Private Ryan" and not see any black soldiers. "We were there," he said. "We did our part."[351]

They certainly did, all of the 320th risking their lives at D-Day, and some losing their lives on those beaches.

In an interview for this book, Dabney said he "felt pretty proud" of being honored. "I wondered if it would ever come in my lifetime." But it also was a bittersweet honor, considering those who were not

---

[349] Linda Hervieux, "All Black Battalion That Landed In Normandy, France, On D-Day to Be Honored On Anniversary Of Siege," *New York Daily News*, June 5, 2009.

[350] Monk interview.

[351] William Dabney, telephone interview, June 9, 2009.

included, he said. "I would have thought they would have recognized us before now. After all, it's been 65 years." [352]

For Dabney, the segregation of Camp Tyson was a little easier to take than it was for Monk and other blacks from the North. "The camp was huge, with African-Americans on one side and whites on the other. I was from Virginia, so I was used to blacks being here and whites being over there. I already was used to colored drinking fountains and all that. It was a matter of, 'We go our way, you go your way.' But the Northern soldiers looked at it differently. It was hard for them to get used to." [353]

There were three all-black units at Camp Tyson, according to Dabney: the 318th, the 319th and the 320th, but it was the 320th that received the combat barrage balloon training. "Actually, the 318th was the one that was supposed to ship out, but they put us on the list first. That's why we ended up at D-Day and they didn't." [354]

Because the 320th was shipped overseas for the D-Day invasion, they received combat training in addition to instruction in how to operate the barrage balloons.

The 320th was unique at D-Day for two reasons: "First, it was the first barrage balloon unit in France and second, it was the first black unit in the segregated American Army to come ashore on D-Day." [355]

NOT SECOND-RATE SOLDIERS

The soldiers in the 320th "were not second-rate soldiers. They were highly trained and took pride in their job. When they were told they were going to land in France to protect the invasion beaches, they quickly realized that the standard VLA (Very Low Altitude) balloon winch was too heavy and cumbersome to lug ashore from a landing craft. The M-1 US Army winch had a gasoline motor and weighed 1,000 pounds. The British Mark VII weighed almost 400 pounds. They developed an expedient by adding two handles to a Signal Corps RL-31 Winder and putting the balloon wire on the DR-4 drum. This new

---

[352] Ibid.
[353] Ibid.
[354] Ibid.
[355] "Sunday Ship History: Behold the Barrage Balloon!" on Eagle Speak web site.

winch weighed only 50 pounds and could easily be carried ashore by one man." [356]

"We were a unit and we always stuck together as a unit," Henry Parham said. "We were considered a special unit and that's why were a little more protected." [357]

But before the men of the 320th were sent overseas, they first had to contend with their experiences at Camp Tyson, not all of which were pleasant for African-American soldiers.

As happened at most U.S. Army camps in that time period, Camp Tyson was not without its share of 'racial outbreaks."

A secret study, commissioned by the U.S. Army, sought to determine whether black troops were really loyal to the American war effort. The study was conducted by Major Bell I. Wiley, who was also an historian from Halls, Tennessee.

Wiley compiled his findings in a 78-page document entitled "The Training of Negro Troops, Study No. 36." The study detailed how the U.S. Armed Forces went about training black soldiers, noncommissioned officers and officers. It also evaluated white officers over black troops, 'riots and racial disturbances' and summarized the lessons learned.

According to the study, "In September 1943, there were minor outbreaks among colored troops at Camp Tyson, Tennessee. Following these incidents, the racial situation in the Army Ground Forces settled down to a long period of comparative calm." [358]

Outlining other, violent incidents at nearby Camp Shelby, Mississippi, and Camp Claiborne, Louisiana, Wiley noted, "Racial disturbances in World War II tended to follow a fairly uniform pattern. Trouble usually began with a real or fancied incident of discrimination or abuse; disaffection was aggravated by circulation of gossip and rumor; a minor incident then precipitated a general outbreak." [359]

---

[356] Ibid, page 8.
[357] "Parham Served In Only Black D-Day Unit." Interview appeared on New Pittsburgh Courier web site, July 7, 2010.
[358] Bell I. Wiley, "The Training of Negro Troops, Study No. 36" (U.S. Army study), page 52.
[359] Ibid.

WIFE SLAPPED, SOLDIER BEATEN

In Southern camps, frustrations were caused when black soldiers did not have access to public transportation and had to find their own way to nearby towns for recreation on weekends. Their access to the same forms of recreation also could be limited.

That was no less true at Camp Tyson. In the days and weeks that led up to the mysterious death of a black soldier from the 320th, there were rumblings of discord between the white and black troops at the camp. Pvt. Herman Hankins was killed by three white Military Policemen in a tourist camp a short walk from the camp on July 31, 1943. (See Chapter 13).

On June 19, 1943, Pvt. Fred Hart of the 319th wrote a letter to Truman Gibson, Civilian Aide to the Secretary of War, informing him of incidents of concern in the camp.

According to Hart's letter, the wife of 2nd Lieutenant Prince Frazier came to camp to visit him and asked to use the phone at Gate 1. She was slapped by the white Military Police on duty. He also wrote that 2nd Lieutenant Evans Taylor was driving in the city of Paris and was stopped by the city police "and was asked, N — , where is your pass...He stated that he was a commissioned officer and he did not need any pass, the results of this he was beaten and put in jail." [360]

Hart also said black soldiers would be waited on last by the white Military Police, even when first in line to get back in camp in the morning.

Hart warned Gibson, "There is a tense feeling there may be a race riot here. Soldiers say that if the officer or Military Police should raise a hand, they would rather die here than on the field of battle. They are tired of being mistreated and seeing colored officers also mistreated. If something doesn't break in the near future, there will be a riot not very long now." [361]

The black soldiers read about the race riots at army camps in Georgia, Mississippi, Texas and California, he continued, "So why not here?" [362]

---

[360] Correspondence from Fred Hart to Truman Gibson, NARA, Camp Tyson, TN. ARC Identifer 633339/MLR Number A1 188 RG 107, Entry 91, Box 186.

[361] Ibid.

[362] Ibid.

There was a "slight belief" by the white officers who were over the black soldiers that there would be trouble, he said. Colonel Leon Reed — one of the white officers in charge of the 320th — had locked up rifles and searched for knives, ammo, and firearms.

Hart also informed Gibson that a black soldier, Dean McCoy, had been arrested for using narcotics (marijuana) and was put in the stockade. "Weight (sic) about 158 pounds when he was taken to the hospital and now he had been beaten up by Military Police in the hospital and weights (sic) about 100 pounds and is not expected to live..." [363]

In a follow-up letter, Hart informed Gibson, "The feeling of a riot here is now gone, there where (sic) about 6 rifles taken, but that is just about all now, so I don't think any thing will happen now." [364]

BALLOON IS SABOTAGED

Early the next month, a protracted investigation of sabotage of a barrage balloon was conducted by the military brass.

The balloon was cut open by a knife at 1 p.m. July 8, at Balloon Site No. 93, while being operated by men in the 318th Barrage Balloon Battalion, Battery C. Several memos are on file in the Library of Congress which relate how extensive the investigation was — the balloon was confiscated, the fabric surrounding it was fingerprinted (but no prints could be obtained), four knives belonging to the men were confiscated and taken to the camp hospital for examination under the microscope.

At least seventeen interviews were conducted of the men in the battery and various motives, including revenge for being reduced in grade or a general desire for sabotage, were considered.

It was noted in the reports that the men already were isolated from the rest of the camp, by virtue of being on duty at a balloon site for training. The usual procedure when training at one of the sites on the outskirts of the camp was that the soldiers would sleep in adjoining barracks to the site and their meals would be brought to them, since they were on 24-hour duty. Balloon Site No. 93 was a mile from headquarters.

In his final report, Tom Blalock, Special Agent with the CIC, determined that there was "deliberate sabotage by one or more crew members" and not sabotage by "any organized groups." From his in-

---

[363] Ibid.
[364] Ibid.

terviews, Blalock had narrowed his suspicions to T/5 George Magness, who the others said had been drinking that day. Until a final determination could be made, Blalock indicated the crew had been restricted to Balloon Site 93 and were under constant surveillance. [365]

BLACK SOLDIERS REPORT INJUSTICES TO BRASS

Meanwhile, letters were being sent from black soldiers to the army brass, informing them that black soldiers were not being picked up at bus stops by the city of Paris bus line and that black soldiers were being either reduced in grade for speaking out or that blacks were being overlooked for promotions altogether.

In an anonymous letter dated July 28, 1943, a soldier wrote that some black officers were relieved of command. The situation, according to the letter, was hurting morale among the black soldiers. "There has been continual Jim-Crow, in the towns, on the buses, at the 'post' theatres, disrespect to officers by white military personnel, without punishment or cessation." [366]

It was three days later that Pvt. Herman Hankins was killed, at which point Pvt. Fred Hart again wrote Gibson. "A colored soldier was killed here by a white MP and they keep it pretty well under cover, to this day, know (sic) one here knows just who it is or in what Battalion the soldier is in but they do know that one was shoot (sic) and where he was shoot (sic), so as to keep down a feeling here they have not told anyone." [367]

With the trauma of the killing of Pvt. Hankins hanging over everyone at the camp, the situation did not improve.

A 'joy ride,' as described in a front page article published by *The Afro-American* newspaper of Baltimore, resulted in sentences of 1-5 years of hard labor for 64 black soldiers from Camp Tyson in December of 1943. [368]

---

[365] Blalock final report, NARA, RG 319, Entry 47, Bx 1319, Ordinance plants, Camp Tyson (folder 004.4), Army Intel Decimal File (1941-1945).

[366] Anonymous letter dated July 28, 1943, NARA, Camp Tyson, TN., ARC Identifier 633339/MLR Number A1, 188 RG 107, Entry 91, Box 186.

[367] Fred Hart letter, Ibid.

[368] "64 Soldiers Get 1-5 Years," *The Afro-American*, Baltimore, Maryland, December 11, 1943, front page.

According to the article, the soldiers "had planned a joy ride to Jackson last October 10, but never reached their destination because camp officers discovered their absence and learned of the missing trucks. They later found the missing men" in the nearby town of Henry.

The sentences "are considered the stiffest ever handed down by an American military court," the article noted. "The non-commissioned officers were given five years each, while one private was sentenced to a year and the other privates received longer terms, according to Col. William H. Dunham, junior commanding officer of the barrage balloon training center."

The men were slated to serve their sentences at a Federal rehabilitation camp, according to a court martial panel, which the newspaper noted was made up entirely of white men.

Eddie Moody of Paris, who worked as a civilian guard at the camp, recalls the incident differently. In his recollection, black soldiers had gone AWOL from the camp and "were on the way to Jackson to start a riot. Everybody on base was worked up about it." [369]

An anonymous letter written months later, on Jan. 5, 1944, informs the army brass that black soldiers still were not being picked up by the city bus line. A handwritten note of May 30, 1944, which apparently was a transcription of an order of Colonel Dodge and Captain W.J. Boggs, relates that Tech. 5 John H. Jones had said, "This is not slavery time" and that following an altercation, he was reduced to the grade of private. [370]

On Feb. 1, 1944, Major Benjamin Green from Headquarters Command had informed 1st Lt. Robert Shamwell of the 319th that he was being relieved from active duty. A letter from Shamwell to Gibson on Feb. 15, 1944, may explain why. Shamwell wrote to Gibson, protesting that black officers were not being given promotions and were being transferred despite high marks. Shamwell noted that because of the situation, "We are now far under strength in officers" and that white officers who were ranked lower than the black officers were being promoted to their positions. [371]

---

[369] Eddie Moody, personal interview with author, May 5, 2009.
[370] Anonymous letter, NARA, Camp Tyson, TN., ARC Identifier, 633339/MLR Number A1 188 RG 107, Entry 91, Box 186.
[371] Shamwell letter, Ibid.

AS IF THEY WERE OURS

Shamwell wrote that the situation was causing the group of officers to be "determined they will not take this lying down even if it means going all the way to the White House in order to get justice done..." [372]

TRAVELING IN GROUPS

In interviews for this book, Monk and Dabney said they generally stayed on base or traveled to nearby larger cities, like Paducah, Memphis or Jackson, when seeking recreation.

With the situation with the city of Paris bus line not picking up black soldiers, they were required to drive themselves to Paris and usually traveled in groups.

"We were told to stay out of Paris," Monk said. "We were told it was terrible as far as blacks were concerned." [373]

In an interview with *USA Today* in May of 1996, the late Corporal Waverly Woodson, who was then 73 and living in Clarksburg, Maryland, said he first encountered racism at Camp Tyson. "The atmosphere was terrible. Everything was totally segregated. The feeling was that blacks were inferior to whites. Most of the officers were white. There were not many black officers, and they only made it to 1st or 2nd Lt." [374]

The late Sergeant George Davison wrote down his experiences at Camp Tyson and D-Day and put them in a box, along with newspaper articles and photographs of the camp that he had saved. His son, Bill, found them after his father had died.

Davison wrote, "Colored troops had it hard back then," but he credited Col. (Leon) Reed of Middlesboro, Ky., who was white and lived with his pregnant wife in Murray, with helping the black troops. "Tops, a hell of a fine fellow, who let no one s — - on his troops at any time. Col. Reed was there, barking my men have the same thing as all other men." [375]

MORALE HIGH?

Perhaps surprisingly, the camp newspaper, *The Gas Bag*, did include columns from the black regiments, just as it did the white units.

---

[372] Ibid.

[373] Monk interview.

[374] "Debt Of Honor," *USA Today*, May 6, 1996.

[375] George Davison notes, property of Bill Davison.

In the April 28, 1943, issue, the article submitted from the 320th, Battery B, reported that "Morale Is High Among 320th Men," noting they were ready for their furloughs and that competition among the platoons was keen during recent field days. [376]

Another article, submitted by the 318th, reports on promotions and feelings of "joy" that furloughs were near. In Battery C, according to Sergeant W.P. Jones, "There isn't much to write about this week, because we are now so settled into the routine of balloon work that there is nothing much new. We just fly old Neoprene Nelly from each site each day from breakfast to taps." [377]

The columns relate light-hearted goings-on among individual members of the 318th, including men who were eager to see their wives on furlough, and plans for an inter-barracks softball game.

In Battery A of the 318th, balloon crew members were proud of high marks they were given for their training. "Every man carried out his assignment with confidence, giving us the feeling that we are ready for the field." [378]

The Gas Bag lists promotions that had been awarded recently, which includes numerous listings from the three black battalions at the camp. Among the promotions announced in the April 28, 1943, issue was that of Aaron Dobbs of Paris, a soldier in the 319th, Headquarters Battery, who was promoted from T-Sgt. To Master Sergeant. [379]

But elsewhere in that edition of *The Gas Bag*, there is a glimmer of the segregation of the black and white battalions: in addition to separate barracks for black and white soldiers, their families also stayed in separate guest houses. The report from Guest House No. 1 lists the families of white soldiers who were visiting, while Guest House No. 2 was occupied by the families of black soldiers. [380]

TRAINING FOR D-DAY

It was against this backdrop that the members of the 318th, 319th and 320th were training in the operation of barrage balloons and were

---

[376] T-5 Charles Taylor, "Reporters Show Morale Is High Among 320th Men," *The Gas Bag*, April 28, 1943, page 4.

[377] Sgt. W. P. Jones, "Furlough Lottery Brings Joy To Those Who Can Produce Enough Lettuce," *The Gas Bag*, April 28, 1943, page 7.

[378] Ibid.

[379] "Promotions," *The Gas Bag*, April 28, 1943, page 12.

[380] "Guest House Notes," *The Gas Bag*, April 18, 1943, page 5.

preparing to be deployed overseas for what would be the ultimate test of their bravery: the D-Day invasion.

Among those who were there, William Dabney, Master Sergeant Wilson Monk Caldwell, and Henry Parham are still alive to tell the tale. Sadly, Sergeant George Davison and Corporal Waverly Woodson are not, although both left their stories behind. All were stationed at Camp Tyson and all were members of the 320th. One of the soldiers among them would later become famous: Bill Pinkney, vocalist with The Drifters singing group, endured the D-Day hardships along with the rest. [381] Two other soldiers from the 320th would later achieve some renown as well: the late George Dennis Leaks, who would become a member of the legendary Dixie Hummingbirds, was in Normandy with the 320th on D-Day [382]. The late James Wilbert Pulley, who later became the chauffeur for Mayor Thomas D'Alesandro and other Baltimore politicians, was a corporal with the 320th. [383]

Unknown to them — or anyone else, for that matter — the United States military was planning a major invasion on the coast of France and had begun preparations for which weapons and which regiments would be utilized for it.

Barrage balloons and the men who knew how to operate them were deemed an important element of the military operation.

To that end, the 320th was sent from Camp Tyson to Camp Shanks, N.Y., then to New York City and boarded ships to cross the Atlantic Ocean to Scotland.

"While we were there, we were constantly training, every day," Monk said. "Then we were sent to southern England, where we were billeted in Abersachen, a little town. Then we went on to Cardiff, Wales, then Southhampton, where we boarded a ship and crossed the English Channel." [384]

The 320th was not told what to expect or what they would be doing, Monk said. "We were just told to get ready, get in line and follow me. I can faintly remember hearing Eisenhower telling us about the invasion." The men were anxious, he said. "There's always a certain amount of fear of the unknown. We certainly weren't going to a tea party." Dabney recalled the intensive training the 320th received in

---

[381] www.blackeyedhandsomeman.blogspot.
[382] Leaks obituary, *The Philadelphia Daily News*, January 23, 1997.
[383] Pulley obituary, *The Baltimore Sun*, July 27, 2003.
[384] Monk interview.

England. "We already were trained on the barrage balloons, but we needed training for combat, for being under fire." [385]

"SOME OF YOU WILL NOT BE RETURNING"

As June drew near and the men boarded the ships to cross the English Channel, Dabney said, "We knew we were going to take part in the invasion. General Eisenhower came on the loud speaker on the ship and said, 'Some of you will not be returning.' That got you upset a bit, hearing that." [386]

Once their ship landed at Omaha Beach, members of Dabney's battery were right behind members of the 29th Texas Rangers. "We weren't the first ones off the ship," he said, "but we were right after the Texas Rangers." [387]

What made Dabney and the other members of the 320th different from the Texas Rangers and other battalions at D-Day was that they were attached to the heavy barrage balloons as they came off the ship. "I had my balloon attached to my belt. I remember flying in the air from the strength of the balloon when we came off the landing board." [388]

Dabney said he weighed 170 pounds at that time "and I had a 25-pound pack on my back" and even with a smaller balloon like he used for the D-Day invasion, the heavy balloons thrust the men in the air. The balloons, he said, were designed "so they wouldn't lift us too high off the ground." [389]

Once on the beach, he said, "just about the time I got off the landing board, the balloon was shot out from under me. I don't know if it was a plane that hit it or if the Germans shot it down, but it was shot down and I immediately disconnected the cable. I threw myself in the sand with my crew." [390]

Dabney and his crew spent two days on the beach, until General Patton and his tank division arrived.

---

[385] Dabney interview.

[386] Ibid.

[387] Ibid.

[388] Ibid.

[389] Ibid.

[390] Ibid.

Monk says now that he doesn't have "a lot of memory of that day. I remember the beach and remember flying the balloons. A lot of the rest is blank and I'm glad because it wasn't a pleasant thing to see." [391]

Parham said in an interview with *The New Pittsburgh Courier*, "You're scared for your life every day. We were waiting for our time to land and when we looked through the glass, it was horrible. When we finally landed there were 15,000 on the beach. You had to watch where you walked." [392]

Parham said he was part of a five-man balloon crew. "We stayed on the beach 68 days. We had to sleep on the ground. So our program was to protect the beach and the equipment and we were very successful." [393]

DEAD ALLIES FLOATING IN THE WATER

In the notes he left behind for his son, Davison wrote that after the decision was made to train the 320th for combat duty, they learned "the rules of ground warfare, kill or be killed, hand-to-hand combat, camouflage, and identification of the enemy."

The soldiers also learned how to fly the combat balloon, which was "a little balloon, about the size of a VW, which operated from a two-man winch with a cable of about 3/8 foot thick and 2000 foot in length. This balloon only took three men to operate while the barrage balloon took eight to ten men." [394]

And Davison explained why the balloons were considered such an important weapon at D-Day.

The balloon had a hitch on the balloon swivel that carried a payload of one pound of TNT "which will get any plane that comes in contact with this cable." [395]

While the balloons are in the air, "planes cannot make a diving bomb run or strafe because they can only see the balloon and not the cable and making contact with the cable means the run is over. There is a crash, loss of plane and pilot. They were very effective." [396]

---

[391] Monk interview.

[392] "Parham Served In Only Black D-Day Unit." Interview appeared on New Pittsburgh Courier web site, July 7, 2010.

[393] Ibid.

[394] George Davison notes.

[395] George Davison notes.

[396] Davison notes.

For Davison, the situation was dicey from the start. Davison's three-man balloon crew was aboard a landing craft that contained 105 mm howitzers and jeeps and even before they reached the beaches, the crew rescued a downed British pilot. When the battle was over and dawn approached, the men could see the bodies of Allied soldiers floating in the water. "It was hell," Davison said. [397]

The U.S. Army's *Stars and Stripes* newspaper covered the 320th's contributions to the invasion at the time. Davison cut out the newspaper's accounts and kept them in his scrapbook.

Noting that the 320th had "the distinction of being the only Negro combat group included in the first assault forces to hit the coasts," the article written by Allen Morrison noted that their balloons "were flown across the channel from hundreds of landing craft, three men to a balloon, and taken ashore under savage fire from enemy batteries." [398]

Some of the men of the 320th "died alongside the infantrymen they came in to protect and their balloons drifted off. But the majority struggled to shore with their balloons and light winches and set up for operation in foxholes on the beach." [399]

According to the article, the battalion's "first kill came recently when a JU88 ran afoul of the cable supporting the balloon commanded by Cpl. George Alston, Norfolk, Va." [400]

The article spotlighted the work of the medics who risked their lives to set up the first aid station on the beach. The men who were praised by the unit's Commanding Officer, Lt. Col. Leon Reed, were "Capt. Robert E. Devitt, Chicago, Illinois; Staff Sergeant Alfred Bell, Memphis Tenn.; Cpl. Waverley B. Woodson, Jr., Philadelphia; Corporal Eugene Worthy, Memphis, Tennessee; and Pfc. Warren W. Capers, Kenbridge, Virginia. All have been recommended for decorations." [401]

---

[397] Jim Moore, "Waynesburg Man Recalls Barrage-Balloon Service," *Observer Reporter*, Washington, Pa, June 9, 1984.

[398] Allan Morrison, "Balloon Umbrella Raised On D-Day Has Sheltered The Beachheads Since D-Day," *Stars and Stripes*, July 5, 1944.

[399] Ibid.

[400] Ibid.

[401] Ibid.

Corporal Woodson, who was interviewed for the book, *We Were There*, had a vivid memory of that day. An Army medic, Woodson was credited with saving hundreds of lives, even after he was injured himself.

"If you ever want to know what hell is like, D-Day was it," Woodson said. "I was on what they call a landing craft, which is a ship that carries troops, vehicles and supplies to the shore. I was in the back of the first wave, and what I saw in front of me was death. All these men, thousands and thousands of men, were heading onto the beach, and the Germans were just shooting them down." [402]

Woodson was injured as a German shell hit the ship his group was in. He was hit in the back and groin, as the landing ship tank he was in floated on the water. Another shell hit the tank and he was thrown into the water.

With debris and bodies in the water, Woodson swam to shore, crawled to a cliff and set up a makeshift first-aid station with a tent roll he found in the water.

"They say I saved three hundred men but I couldn't tell you how many," Woodson said. "The newspaper articles I got here says I worked thirty hours after being hit, but I can't remember. I just know it was a long time." [403]

In the photographs of the D-Day invasion, some of the most iconic images are the barrage balloons that are seen floating in the air, amid the smoke and the battleships along the shoreline.

And yet, the men who manned the balloons were largely forgotten.

Two videos of the men of the 320th on D-Day have recently been uncovered and can be seen on YouTube today. One shows the men in a landing craft on the way to the battle. They are pensive, quiet and some sit quietly, reading French dictionaries. The other shows the aftermath of the battle, as an integrated memorial service for the fallen is held on the beach.

---

[402] Yvonne Latty, *We Were There* (Harper Collins, 2004), page 29
[403] Ibid, page 30.

Woodson received a Bronze Star for heroism and a Purple Heart. "Four or five of us were recommended for higher honors, but we didn't get them," he told USA Today. "Our white superiors recommended us but the War Department didn't approve it. Silver Stars and Congressional Medals of Honor did not go to black officers or black enlisted men." [404]

On the fiftieth anniversary of D-Day, Woodson was one of three soldiers who were honored with a ceremony on Omaha Beach. "I don't know why they chose me," he said, "but it was a wonderful thing. I was the only black man of the three. I think it was the French's way of saying, 'Thanks.'" [405]

Dabney was honored with the Legion of Honor in 2009.

Davison received service medals, including for good conduct. But, his son, Bill, said, "He didn't receive anything specific for the job he did at D-Day. There was nothing for the men who participated in the invasion on the beach and he really felt there should have been." [406]

---

[404] "Debt Of Honor," *USA Today*, May 6, 1996.

[405] *We Were There*, page 31.

[406] Bill Davison, telephone interview, September 8, 2009.

Chapter 16

Working There Was
An Education In Itself

I T is no exaggeration to say that most everyone in Henry County owed their livelihood to Camp Tyson, either directly: from increased business at their shops or companies; or indirectly: through the side businesses that developed because of the camp.

Hundreds of civilians worked at the camp in a variety of capacities. They helped to keep the Camp operating like a well-oiled machine as gate keepers, guards, firemen, secretaries, laundry workers, mechanics, clerks, soda fountain operators and waiters, just to name a few.

For most, it was a positive and memorable experience. For a select few — especially those who worked in the huge, hot laundry — not so much.

"IT WAS ENJOYABLE, EXCITING, SOMETHING DIFFERENT"

For Jeanne Anderson Townsend, it was a chance of a lifetime, one that took her directly into the inner-workings of the camp. Her experience at the former Toler Business College in Paris, which she attended right after high school, helped her land a job at the camp, eventually working as secretary to a few of the officers.

"The camp needed people to work," she said. "Everyone could have a job that wanted one. They put a lot of people to work who hadn't had a job in a long time." [407]

But her job was more specialized, requiring an interview and passage of a federal civil service exam. Her first job at the camp was working at headquarters, cutting stencils and running copies. "I did that for a month, then they asked me to work in the ordinance office, then the barrage center, then the station complement, and then as secretary at the motor pool." [408]

She worked for Battery C of the 302nd and in the motor pool, taking dictation, typing up orders for the people that worked in the mo-

---

[407] Jeanne Townsend, personal interview with author, June 2, 2009.
[408] Ibid.

tor pool, and taking shorthand from conversations she was told to listen to over the phone. "That was interesting," Townsend said. "Every now and then I was asked to listen on the phone while the officers were speaking to someone and take notes from their conversations. Did the people on the other side of the line know someone was listening? No, not really." [409]

Townsend took the bus operated by Blake Bus Lines to and from work. "They had several bus stops set up around town and I would stand at the bus stop at Blakemore and Wood and get the bus there," she said. Once at the camp, security was tight. "When you pulled up at the gate, the Military Police were there and you had to present your ID badge." [410]

Being in the thick of things, Townsend was able to experience being around the German prisoners of war when they made a cradle for her boss' new baby and sent a cake they made in the bakery to her office (See Chapter 23).

She also was able to meet a lot of different types of people. "In our office, we had a lot of enlisted men from places like Poland, Switzerland, Lebanon. Working there was an education in itself. I really learned a lot," she said. "It was enjoyable, exciting, something different." [411]

Among Townsend's most treasured possessions are photographs of herself, her co-workers and her bosses, Maj. Phillip G. Kinken of North Carolina and Capt. William Sanders, Atlanta, and a still-intact invitation to a 302nd Barrage Balloon Battalion informal dance at Paris city auditorium on March 7, 1942.

"Those really were good times," she said. [412]

ALWAYS SOMEONE COMING AND GOING

The late Eddie Moody of Paris worked as an auxiliary MP, truck driver and chauffeur for General John Maynard at the camp, and for a young man his experience there was fraught with excitement. So much so that Moody made it his lifelong quest that Camp Tyson not be forgotten, organizing soldiers' reunions over the years and collecting as much memorabilia on the camp as he can.

---

[409] Ibid.
[410] Ibid.
[411] Ibid.
[412] Ibid.

Moody enlisted in the Navy and was on a work detail when he was injured. He was discharged after five months, but still wanted to contribute to his country. "We already had two boys in our family in the Navy and my parents really didn't want me to enlist," he said. "But I wanted to do something, so I asked for a job at the camp." [413]

As civil service employees, Moody said, "We were called 'citizen soldiers' and I took it seriously." He said he worked at the gate, where security was tight. "When someone came to the gate, we would look at their face, look at the picture on their badge and if they matched, we'd let them in," he said. [414]

His favorite duty was chauffeuring Gen. Maynard. "He would use me to bring him in to town because I was local. He ran that camp like a well-oiled machine and I would protect him any way I could." [415]

Every working day, employees would go to the bulletin board "to see what your duties were and it would have the 'word of the day' on the board — the password that you'd be using that day. You'd go to work, wherever the officers would tell you to go," Moody said. Single soldiers were paid $21 a month and if they were married, $50 a month, he said. "I remember that the married soldiers would be so proud to get in to town to see their wives," he said. "And at 5 p.m., you'd see the wives at the bus stops waiting for the husbands to come home." [416]

For Moody, it was an exciting time. "There was always someone coming and going." [417]

"THE CLOTHES JUST CAME IN A BIG HAMPER"

Faye Davis Kesterson had one of the more interesting jobs at the camp: working for the quartermaster's corps sizing all the soldiers' uniforms.

A newlywed in need of a job at the age of 18, she did what most people did then: looked for a job at Camp Tyson. "They put me in with the other ladies who were sizing the uniforms," she said. "I remember we were in a huge warehouse. There were just tables and chairs in there and it was sort of like an assembly line. The clothes just came in a big hamper and they would put them down on the table and

---

[413] Eddie Moody, personal interview with author, June 5, 2009.
[414] Ibid.
[415] Ibid.
[416] Ibid.
[417] Ibid.

we would measure each piece of clothing, the length, width, and then put the sizes on the front."

Kesterson and the group she worked with "just did the shirts. I assume someone else did pants. We did each one by hand. There was no machinery in the building. All of us who worked in there were women, but soldiers would come in and bring the big containers full of the clothes because they were too big for us to handle. We worked five days a week and we were always busy." [418]

"I Missed Him Terribly"

For Norma Clayton of Paris, it was a bittersweet experience born of necessity.

She already was married to Kenneth Clayton, who was stationed in the Pacific, near Australia. Needing a job while he was gone, she started working as a clerical employee at the camp. "I was making do," she said. "I missed him terribly." [419]

She and several other civilian employees would carpool to and from work every day. "George Looney would drive and we'd meet at the post office in Paris. Gas was rationed, so we had to car pool. It wasn't a bad experience. At least we all were compatible." [420]

For her, "it wasn't really a happy time, but I did meet a lot of people out there. I was just wanting Ken to come home and wasn't thinking of much else," she said. [421]

"Just Lucky to Get that Job"

The late Jimmy Huffman of Paris lived "way out in the country, out in Mansfield," a small town several miles from the camp, when he landed his job there. "Nearly everyone worked out there and I was just lucky to get it because everyone in the county was applying for those jobs." [422]

The job he acquired was a particularly interesting one, as time-keeper for camp employees. "Every employee would check in and get their badge when they'd arrive for work and then come back to our

---

[418] Faye Kesterson, personal interview with author, November 9, 2010.
[419] Norma Clayton, personal interview with author, January 13, 2010.
[420] Ibid.
[421] Ibid.
[422] Jimmy Huffman, personal interview with author, April 16, 2009.

shack when it was quitting time to turn their badge in," he said. Three timekeepers worked in his shack "and most of the laborers were assigned to come through my shack," he said. "There were so many people out there you didn't see the same person two days in a row." [423]

Everyone wanted to work at the camp, where good-paying jobs could be had. "It was the end of the Depression and people around here couldn't find work. Carpenters had been paid 40 cents an hour before but out at the camp, they could get $1.10. That's a good increase." It could be tedious work, he said. "We worked seven days a week, 7 a.m. to 5 p.m. and we very seldom saw daylight."

A former Civilian Conservation Corps worker during the Depression, Huffman's wartime experiences after Camp Tyson were far less tedious. A member of the 4th Armored Division, he served 18 months in the Army, serving under General George S. Patton. "We used to see Patton a lot," he said. "He rode ahead of us, shooting at Germans with a rifle." [424]

A NEED FOR FIREMEN

In a huge Army camp made of wood and with plenty of combustible materials, one of the first things you need is a fire department. A call went out from the camp for civilian firefighters. First one fire house was built on site and as the camp grew, a second one was constructed on the other side of the camp.

Two local women whose fathers worked as firemen at the camp are Nelda (Smith) Pinson and Brenda (Argo) Lewis.

Pinson's father, Rexie Smith, was employed as one of the first camp firemen shortly after the first firehouse was built. The Smiths lived in the small community of VanDyke in Henry County, and Rexie and another local man, Bill Hart, used to travel to work at the firehouse together.

For Smith, it became a lifelong passion. His time at Camp Tyson was interrupted when he was called to serve in the Navy aboard the USS Eastland in the Philippines in 1944. But after he returned stateside, his daughter said, "Firefighting became his career." [425]

Brenda Lewis' father, William Argo, was already a fireman in McKenzie in nearby Carroll Co. when he began work at Camp Tyson.

---

[423] Ibid.

[424] Ibid.

[425] Nelda Pinson, personal interview with author, October 9, 2009.

"He was a civilian employee, employed by the War Department as Civil Service," Lewis said. For Argo, his tenure at Camp Tyson was long, from July 1, 1942-Jan. 5, 1947, way after the camp closed and he was furloughed for "lack of work." [426]

Lewis now lives in Huntingdon and owns several pieces of memorabilia that belonged to her father, including his red fireman's cap, which is on display at the Gordon Browning Museum in McKenzie.

Jerry Ridgeway of Paris often looks at photographs of her mother taken while she was an employee at the camp. On one photograph her mother inscribed on the bottom, "Edith's friends," who included construction workers and soldiers who she got to know while working at the camp.

Her mother, whose maiden name was Edith Moser, worked for General Maynard while the camp was still under construction. For her mother, she said, it was an exciting time. "The place was supercharged," Ridgeway said. "Anyone who could, did anything to help the camp. It was a patriotic time." [427]

WHEN YOU HAVE SOLDIERS, YOU HAVE DIRTY LAUNDRY

When there are thousands of soldiers, there is a real need for a laundry.

The laundry at Camp Tyson was a huge, industrial-sized affair, with dozens of hot, steamy machines in operation full-bore all the time.

It could be a tedious job and for others, downright miserable.

But for J.D. Wilson of Paris, it was "a nice place to work. It was terrific. We washed the khakis, underwear and socks. I worked in the laundry for two years until I was drafted into the Marines." Wilson remembers the experience as exciting, with "a lot of people coming and going all the time." [428]

Marie (Holley) Donaldson, who now lives in Knoxville, worked in the laundry as a teenager. "I was 15 and there were three or four girls and I who went out there looking for a summer job. I got a job in the

---

[426] Brenda Lewis, email correspondence, June 4, 2009.
[427] Jerry Ridgeway, personal interview with author, May 1, 2009.
[428] J.D. Wilson, telephone interview with author, May 20, 2009.

laundry. The soldiers would come in with their bags full of dirty clothes and I would sort them." [429]

She remembers catching the bus to the Camp from Paris every day with other young girls. "We would snicker at the soldiers, like girls do. We acted so immature," she said. [430]

When the bus would pull in, she said, "the laundry was off to the right and we'd walk right to it, but we had to pass the POW section. Sometimes, when we'd go out for lunch, they'd be standing outside, over the fence, and they'd whistle. We weren't allowed to talk to them and they'd get in trouble when they whistled at us." [431]

After her summer experience at Camp Tyson, Donaldson got a much more stressful job — at the Oak Ridge Nuclear Plant in Oak Ridge, Tennessee.

"It Was Hot as Blue Blazes"

The laundry experience was far less pleasant for Ethel Wheat of Henry. She was 16 or 17 and rode a bus from McKenzie, working at the camp for a year. She worked pressing shirts in the laundry, which was a rather arduous task.

"The laundry was a big old building, bigger than any barracks," she said. "Those soldier boys would bring the clothes in in a big bag and then they'd put them down a chute. The clothes would be washed and I would press them." To press the clothes, she said, "You'd put them on something, I can't remember what you called it, and the arms of their shirts would stick out. We'd iron down the collar and the cuffs, very professionally. Of course, it couldn't be any other way, since the inspectors would come by and it had to be perfect." [432]

For Wheat, the whole experience was rather unpleasant. "I really don't remember seeing any barrage balloons and didn't see many soldier boys, either. I stayed in that building most of the time. I ate lunch right at my machine. I brought a sandwich with me every day. We worked from 7 a.m. to 4 p.m. every day, five days a week." [433]

---

[429] Marie Holley Donaldson, telephone interview with author, August 5, 2009.

[430] Ibid.

[431] Ibid.

[432] Ethel Wheat, telephone interview with author, September 9, 2009.

[433] Ibid.

Wheat remembers the soldiers' laundry went through the process of "washing, ironing, finishing, folding and wrapping up real nice in paper for them to pick up." There were hundreds of machines in the building, with "big old round fans that pulled the air out. If it got too hot from pressing, you were welcome to bring you a little fan and put it on the table and help you stay cool that way," she said. [434]

Both men and women worked in the laundry and were kept busy. "Those soldier boys brought laundry in there every day. Sheets, pillow cases, blankets, uniforms, everything they had. And they wanted everything to look crisp and sharp." The men would help the women pull the heavy clothes out of the washers, she said. "And the guards would watch you. They didn't want any hanky-panky, no talking to soldiers. But I wasn't interested in that anyway. I just wanted a job." [435]

But for Wheat, the experience "wasn't that exciting. It was hot as blue blazes in that building. And it was hard work." [436]

---

[434] Ibid.
[435] Ibid.
[436] Ibid.

As If They Were Ours

Chapter 17

A Close Encounter
Of A General Kind

AT 16 years old, Betty (Alexander) Brewer had landed a summer job at Camp Tyson — working in the soda fountain at the Post Exchange (PX) there.

"It was the summer between my junior and senior years," Brewer said. "And I made sodas. It was actually a lot of fun."

It was rather a family affair for Brewer at the camp, since her father was a civilian employee in the maintenance department and her sister worked as a bookkeeper at the PX.

"We were fairly busy," she recalled. "There were a lot of soldiers in and out. I made sodas, dipped ice cream. When the soldiers were hot from marching in the mornings, they'd come in and get something: a soda or ice cream. They treated me like a kid, which I was, and would joke around with me a lot."

Summer in Tennessee can be very hot and one morning, things got a bit hotter for Brewer in the PX.

"My boss at the PX, Major Cobb, had told us not to let soldiers walk through the storeroom as a short cut because some things had come up missing," Brewer said. "I was told, 'If you see anyone back here, don't allow it. Tell them to leave.'"

Brewer was busy in the storeroom, readying containers of fudge one morning. Out of the corner of her eye, she saw a man in uniform starting to walk through. "I said, 'You're not supposed to be in here.' I guess pretty rudely, and he left. I didn't think anything about it and went about my work."

A few minutes later, Major Cobb came into the PX with the same gentleman in tow. "He came up to me and said, 'I'd like you to meet General Maynard and he can go anywhere he pleases.'"

Brewer said, "The General almost had a smile on his face but I'm sure he thought I was an upstart of a girl for telling him to get out. I remember that Major Cobb was pretty nervous. I was just doing what Major Cobb told us to do, but I guess I did my job a little too well that day."

That was not Brewer's only uncomfortable encounter with an officer at the camp.

"There was a separate room for the officers and we'd go in there to take their orders and then carry it over to them," Brewer said. "I remember one day I spilled a milk shake all over a 2ⁿᵈ Lieutenant. I just stood there, watching it ooze down his sleeve."

Brewer grabbed a towel to help him clean up "and he said, 'No, thank you. I'll handle it.'"

The officer took it well, she said. "After that every time he'd come in he'd tell the others, 'Watch out for her. She's dangerous.'" [437]

---

[437] Betty Brewer, telephone interview with author, July 20, 2009.

AS IF THEY WERE OURS

Chapter 18

"This Is Your Life, Pearl Routon"

PEARL Routon, who some called "The Queen Bee of Camp Tyson," was a firebrand.

From the outside, Routon appeared to be just what she was — a sweet lady who had a passion for flowers. Upon closer inspection, however, you would find an astute businesswoman, renowned artist, floral designer, and one-woman public relations machine.

She seemed destined to play a central role in the history of Camp Tyson from the very beginning. After all, it was her husband's family which lent its name to the small town of Routon, where the camp was located.

Routon and her husband, Joe, owned most of the land across the road from Camp Tyson and, as her granddaughter, Stephanie Routon Tayloe, put it, "She always had an eye to make a dollar, so when Camp Tyson was established she seized the opportunity." [438]

Seizing opportunities was Pearl's trademark and she began by envisioning what the soldiers and their families would need most: Housing, food, good fellowship.

She constructed two rows of cabins, furnished them with the basics, and she converted the family farm's old tenant houses into apartments, renting both to enlisted men and their wives.'

Pearl built a café which she called "The Dinner Bell" and included a room for a local post office, after which she was able to have herself appointed postmistress. She also built a small country store. These were reportedly the only businesses in the small town of Routon during this period — except for a honky-tonk called "The Trees" which was not owned by the Routons.

A lifelong booster for the local area and a mother of three soldiers herself, she applied her skills of persuasion toward urging other families in Paris to rent space to the military families. Joe and Pearl Routon lived with their family in a huge Antebellum home on Dunlap Street in Paris and they surrounded themselves with soldiers there, too, rent-

---

[438] Stephanie Tayloe, "Miss Pearl Routon During Camp Tyson Days," property of Stephanie Tayloe.

ing out the upstairs, basement and an adjoining log cabin to various families.

Once Camp Tyson opened and as World War II became more intense, Pearl devoted herself to the war effort. She operated a flower shop and greenhouse, but "during the worst years of the war, she pretty much left the florist business to Daddy Jim...while she spent her days at the Dinner Bell and the post office in Routon," according to her granddaughter. [439]

As postmistress, Pearl's duties were varied. Before the depot at Camp Tyson was built, she was called upon to play a central role in ensuring that the mail was delivered from Routon. With two trains passing through Routon every day, an arrangement was made for Pearl "to hang the sacks of mail on a telephone pole by the railroad and the train would slow down enough for a trainman to grab the sack of mail with a pole and he would then toss out a big bag of incoming mail for her to put in the boxes at their post office," Tayloe said. "During the war, the mail in Routon was very, very heavy, more than the mail at Paris."

Pearl's daughter-in-law, Val Routon, recalled that "The Dinner Bell" was outfitted with a juke box with "records of big bands and patriotic music. Breakfast featured home-made biscuits and lunch was a plate lunch, soup and sandwiches." [440]

The cabins were furnished with "the bare essentials," Val Routon said. "It was one big room, with a hot plate for cooking and a bed. It was like what you used to have at tourist camps. But she had a waiting list of people wanting to live there." [441]

The cook at "The Dinner Bell" was "a black lady named Genie who had been a well-known cook for several Paris cafes before the war," she said. "She had these huge wood-burning stoves going all the time, with big pots of beans, turnip greens. Roasts and meat loafs in the oven," Val Routon said. "The Dinner Bell" was a busy place. "There were always masses of people there. The civilian employees from the camp would eat there and you'd get the people from Paris who would come on Sundays, too. [442]

---

[439] Ibid.
[440] Val Routon, personal interview with author, September 10, 2009.
[441] Ibid.
[442] Ibid.

Pearl Routon became friends with General John B. Maynard, the commander at the camp, and always made sure he had "a special steak" when he dropped by the restaurant, according to Val. [443]

While Val Routon's husband was in the service, she spent most of her days with Pearl in Routon, she said, "and I can remember the awesome feeling of watching those huge barrage balloons floating in the air at Routon over the camp, thinking that they were designed to be 'weapons' against the enemy and that the soldiers who would implement them in actual warfare were right here in Henry County." [444]

With the war over and all her sons home safely from the war, Pearl could relax. She and her husband continued operating the florist business and Pearl honed her artistic talents. Specializing in oil paintings, she also became a proficient china painter. Her portraits have hung in the Governor's Mansion over the years.

Routon put her powers of persuasion to good use once again when she successfully lobbied the State of Tennessee to have her signature flower, the iris, named the state flower.

In 1956, Pearl was told by friends that she was needed in Hollywood to see a celebrity about getting a commission for a painting. While there, she was in the audience for a taping of Ralph Edwards' "This Is Your Life" program.

The camera panned on her face as Edwards announced who the surprise subject of that night's show would be: "Pearl Routon, This Is Your Life," he said, and Pearl for a moment sat in stunned silence and then began smiling and laughing.

All of Paris watched the broadcast of the show when it appeared on the air a few days later. A parade of people from her past and present, including Tennessee Gov. Frank Clement, came on the stage, each bearing a rose for Pearl.

Among the guests was one of the biggest surprises for her — Tom Joy, who was a former Camp Tyson soldier who had lived in the log cabin adjoining the Routon's home on Dunlap Street.

Joy spoke of the generosity showed to him and his wife by the Routons during the Camp Tyson days and related how much of an influence Pearl was on him. So much of an influence, in fact, that Joy later opened his own flower shop in Nashville. [445]

---

[443] Ibid.

[444] Ibid.

[445] "Pearl Routon," *This is Your Life*, videocassette, property of Stephanie Tayloe.

As a prize for appearing on the program, Edwards announced that when she returned to Paris, there would be a gleaming new sign at her flower shop. She also was given a new 1957 Studebaker in which to tool around town and a movie projector on which to watch the film of the television show in the privacy of her own home.

Stephanie Routon still has the film of the show and watches the program periodically.

TWO COINCIDENCES

"I Remember the Whole Works"

James Guy, 76, of Sevierville, Tenn., was driving a taxi cab one day when a couple from Paris entered his cab. "I asked them where they were from and they said Paris, Tennessee. I told them my family lived in Routon during the war and had worked for Pearl Routon." Phone numbers were exchanged and Guy eventually called one of Pearl's sons, Richard, to go over old times.

"I was only four years old at the time," Guy said, "but I remember a lot about it. I remember everything about how it looked. I remember the restaurant, the grocery store, the post office. We lived behind that, in a little house." [446]

Guy said his father, Lacey Guy, maintained the area for Pearl. "He took care of everything, the gas station, the post office. My Dad built bridges before the war and how he became acquainted with Pearl, I don't know." [447]

"I remember the whole works. Soldiers lived in rental places, small apartments," he said. "And I can remember going down the road to the camp to watch the dirigibles. The camp was a busy place." [448]

"I Didn't Even Know He Had Been Stationed Here"

On business, Tom Little had been transferred to Henry County. Speaking with his mother, Helen, he learned that his father, John, had been a soldier at Camp Tyson in Routon. "I had never heard of Camp

---

[446] James Guy, telephone interview with author, November 18, 2009.
[447] Ibid.
[448] Ibid.

Tyson and I didn't even know he had been stationed here," Little said. "It was just the strangest coincidence that I should end up here, too." [449]

John Little of St. Louis had been in Quartermaster training at Camp Tyson (in the 476 QM Truck Regiment Co. L) and while there, he and his wife were seeking housing. They found a place to live, in the basement at 902 Dunlap Street — the home of Joe and Pearl Routon.

That information was a surprise to Stephanie Tayloe, as well. "I never knew anyone had lived in the basement. I can't believe that someone could even live down there," she said. "But I guess knowing Pearl like I did; I guess I can believe that she could stretch every bit of space in the house for the soldiers." [450]

Little and Tayloe had a good conversation at the library where she works and he was able to share with her what happened to his parents after they left the Routon home.

After his Camp Tyson truck training, John Little was first sent to England and then to France and was among the soldiers in the first wave at D-Day. But on June 18 after the Allied invasion, Little was driving a jeep in Normandy, which Little said "got in the way of a German tank." [451]

Little was shot in the leg, taken prisoner and sent to a Berlin hospital under German guard. Medical attention was lacking there, Little said, and the bullet was not removed. [452]

Little still has the telegrams that were sent to his mother after his father was taken prisoner. The first communication, dated July 14, 1944, regretfully informs her that her husband was missing in action since July 18 in France.

Further communication, in February 1945, informs her that he was being held in Stalag 3C in Germany, and then came the good news, sent by telegram in March, that her husband was being released.

Little still has the newspaper articles from *The St. Louis Post-Dispatch* with photographs showing his father's release and homecoming in St. Louis.

His mother sent him some other mementoes from his war years, too: two tickets to the Service Club at Camp Tyson, price 25 cents; a temporary pass for John Little to visit his wife in Paris on November

---

[449] Tom Little, personal interview with author, August 4, 2009.
[450] Stephanie Tayloe, personal interview with author, September 10, 2009.
[451] Tom Little interview.
[452] Ibid.

16, 1943; and a pass for Helen Little, 902 Dunlap St., Paris, to visit her husband at the camp on Nov. 18.

And, Little has made sure to drive past the huge Antebellum home on Dunlap St. more than once to see the spot where the Routons made his parents welcome during the war.

Chapter 19

4,000 Hamburgers To Go

THE railroads, already the main mode of travel across America, became an absolute necessity during World War II.

In addition to the copious passenger trains, the railroads were essential to the war effort by transporting troops, war supplies and materials. With two major rail lines coming through Paris — the L&N (Louisville and Nashville) and the NC and St. L (North Carolina and St. Louis) — and Camp Tyson located right outside of town, the railroads were basic to daily life for Henry County and its population.

But it was the L&N Railroad that played a central role for Camp Tyson.

Four months before the attack on Pearl Harbor and as the camp itself was being built, L&N track crews were rushing to complete another spur track to connect the Memphis Line to the construction site which would become Camp Tyson. The five-mile spur was completed at a cost of $48,000 in August of 1941. [453]

After the Pearl Harbor attack, the camp was rushed to completion, with the first barrage balloon launched there on February 13, 1942.

Soon, passengers on the Memphis Line trains "began seeing dozens of large airborne balloons while passing the Routon countryside." And it became something for passengers to look forward to. "Throughout the war, the barrage balloon spectacle interested L&N's passengers." [454]

The passengers aboard the line in the spring of 1943 had a special treat when 165 balloons were aloft in the skies over Camp Tyson for a photographic exhibition. [455]

The 3.25 miles of track inside Camp Tyson was maintained by the U.S. Army and laid with a 120-pound rail. According to Dennis R. Mize, who wrote a book on the L&N called *The L&N Memphis Line*, the camp was switched by a fifty-ton U.S. Army 0-6-0 switcher No. 5017, which was nicknamed "Susie." [456]

---

[453] Dennis R. Mize, *L&N's Memphis Line* (Port Charlotte, Florida: MFS Line Publishing, 1999) page 145.

[454] Ibid.

[455] Ibid.

[456] Mize, 146.

And Camp Tyson was a busy place. Some 6,500 tons of freight was moved by rail each month once the camp opened. "This amount of tonnage required prompt loading and unloading," according to Mize. "On one occasion, eighty-five empty boxcars and flatcars were placed during the morning for loading. Ten hours later, all eighty-five loads were placed back on the interchange track!"[457]

NEW DEPOT IS ESTABLISHED

To handle the traffic, the L&N established a new station agency, named Camp Tyson. L&N Agent J.R. Redmond was sent from Henry, just a few miles from the camp, to run the agency while the camp was under construction.

At first, Redmond worked out of a set-off boxcar. "He found himself inundated when 'hundreds of loaded cars descended upon the station each week. Mr. Redmond had to keep stepping to keep his head above water and to prevent traffic from becoming hopelessly snarled.'" Eventually, two clerks were assigned to help Redmond: chief clerk W.H. Guthrie and clerk J.G. Tosh.[458]

Likewise, passenger traffic increased as well and the L&N designed and constructed a new depot at the camp. A rush was put on the job and the depot was completed March 14, 1942. "Built at a cost of $5,500, the frame-construction depot was 35' by 52.' A 14' shed was built in front of the depot, as well as an 800-foot screenings' platform," according to Mize.[459]

The Camp Tyson depot became a major passenger stop for military personnel, according to Mize. The Pan American train stopped for passenger traffic "going to, from, or beyond" Memphis or Guthrie, Ky. Two daytime "locals" also stopped on signal at Camp Tyson. "At any time during the war, several barrage balloon battalions consisting of about 1,150 men each, were in training at the camp" and each battalion arrived and departed by train.[460]

Gasoline and tire rationing during the war made it difficult, if not impossible, for people to travel great distances by automobile. Railroads, therefore, became the main form of travel.

---

[457] Ibid.

[458] Ibid.

[459] Ibid.

[460] Ibid.

Train depots became especially busy places, with station agents stretched to the limit, and the local depots in Paris were no exception.

WATCHING THE WORLD GO BY

Jeannette Snow McFarlin remembers a wartime activity for her family was to drive from their Park Street home in Paris to the L&N depot on Depot St. and watch the comings and goings there. "It was actually fun. They had a lot of benches and chairs in the depot and outside and we'd sit and watch people coming and going from the trains. And we weren't the only people in Paris who did that. It was a busy place and it was interesting to watch all the activity." [461]

Passenger trains were so busy during the war that it was not uncommon for passengers to ride in the engine cab because there was no more standing room in the cars. "Like many railroads, the L&N was forced to resort to ads in local newspapers discouraging non-essential travel." [462]

The Paris L&N depot was huge. An elaborate and picturesque two-story frame building, the freight and ticket agency, along with waiting rooms, were on the first floor. On the second floor were the employee offices, including yardmaster, car distributor, trainmaster, superintendent, dispatchers and clerks.

Freight business also mushroomed, with local industries, such as the Salant & Salant shirt factory in Paris having wartime production contracts.

F. Paschall Key recalled the effect that the troop trains had on Dinwiddie's Hamburger Stand in Paris. "One day, early in the war, George Dinwiddie received an unbelievable order for 4,000 hamburgers to go: in four hours, when a troop train was due. Dinwiddie bought up all the ground beef in town, every paper bag he could get, and every packing box. He managed to deliver the order on time at the (rail) yards just as the train pulled in. Similar orders occurred several more times, and Dinwiddie was better prepared the next time." [463]

Dinwiddie was not the only person who had experience with providing large amounts of food for troop trains. Julia Hutson of Paris was able to maintain a profitable — and busy — side business supplying sandwiches for the troop trains which passed through town.

---

[461] Jeannette Snow McFarlin, personal interview with author, July 22, 2010.

[462] Mize, page 83.

[463] Mize, page 148.

They Started Asking Us Girls to Work

Jewell Phelps had a front row view of just how busy the railroads were when the camp was open. At one time or another, Phelps worked at each stop on the L&N Line, from Arlington, Bells, Brownsville, Humboldt, Erin, Clarksville and all the little stops in between.

"So many of our boys were in the war, they started asking us girls to work," Phelps said. She worked as a dispatcher, "keeping up with where the trains were going," as well as handling the packages that were delivered. "The trains were full then," she said. "Soldiers were going back and forth to their families and families going to different towns to visit them. People were standing in the aisles all the times. It was a very exciting time; a good time to meet a lot of interesting people."[464]

Sack Lunches for Soldiers

Julia Hutson recalls that the government paid her to make lunch bags for the troop trains: as many as 125 at a time. Typically, she said, she would put in each paper bag a sandwich (ham, cheese, bologna or pimiento), cupcakes or cookies, apple or banana and carry them to the depot from her Edgewood Street home in Paris. [465]

Key also recalled that the Memphis Line also was used to transport German POWs. He recalled seeing special POW trains filled with German prisoners on the way to internment camps in Arkansas and that they regularly stopped at the Paris depot for food. "The food was always taken into the cars to be served to the prisoners, who were kept under military guard." [466]

The Camp Tyson depot was torn down after the war, but the spur that was built to connect the camp to Paris is still used. It was used by the Spinks Clay Co., which was constructed on the Camp Tyson site after the camp was discontinued and it still is used by the Belgian-owned company which took over the Spinks operation in recent years.

---

[464] Jewell Phelps, personal interview with author, November 19, 2010.

[465] Julia Hutson, personal interview with author, August 18, 2009.

[466] Mize, page. 148.

　　　　　　　　　　　　　　　　　　As If They Were Ours

Chapter 20

"Paris Has Nowhere To Go But Up, So I'm Staying Right Here"

WITH all the raw emotion, anxiety, and sheer excitement, war can be a perfect backdrop for romance.

Camp Tyson did not disappoint, with numerous marriages spawned there. Big-city Northerners who never would have thought they would end up living in the rural South did just that because they fell in love and married local girls.

Yes, several of the couples who began their married lives at Camp Tyson did move on to other places, but most settled right in Henry and Carroll County. And the unions of the boys from far away and their local girls helped the local area grow even more after the Camp closed.

And it is remarkable how many of the soldiers knew instantly the girls they would marry were "the ones" upon their first encounters. From Aaron Dobbs, Sr. to Cal Orris to Kenneth Bersey, the spark of lifelong love was kindled on first sight.

EATING EASTER DINNER AND STAYING FOR A LIFETIME

A group of soldiers appeared one day in the field abutting the sprawling family farm where Winnie Perkins grew up.

They had been working in the field, which was in the outer reaches of Camp Tyson, working with the barrage balloons and they had a little campsite set up for themselves.

It was nearing Easter Sunday in 1943, they were homesick and they were tired of eating the chow served at the camp. They saw a farmhouse and walked up to it. Finding the lady of the house, Addie Greer, they boldly asked if she would bake them an Easter dinner they could take back to the camp. She told them she would and then directed them to another nearby house, where she said they would help, too. "And she told them that they have a pretty daughter over there, too," according to Winnie Perkins — the very same 'pretty daughter' to whom Mrs. Greer was referring. [467]

---

[467] Winnie Perkins, personal interview with author, July 22, 2009.

The neighbor ladies made several dishes for the soldiers, who came to pick them up to take back to the camp for their big feast. In the process, Bill Perkins took notice of Winnie. "Bill came back the next day," Winnie said, "and he kept coming back. It always seemed he would come at dinner time, always with an excuse. He said he was hungry for good country cooking." [468]

Bill asked her on a date and she accepted. But she maintained a dating life with other boys as well. "I still lived at home, but I had my own job, working as a secretary for a law firm, and I wasn't a child anymore. I didn't have a lot of time for a steady." But the soldier from Virginia was persistent. "Bill was very insistent. He just kept calling back and coming over. I reckon he liked me because I was a farm girl and he was a farm boy." [469]

Their dates were simple, going to the movies on base or staying at her home, popping popcorn, playing cards. When they went to town, Bill joked, "We would split a coke." After Winnie met "and fell in love" with Bill's family, the two were married, and Winnie told him they could get an apartment in Paris. "He made it clear he didn't want that. He wanted to live in the country. I said, 'OK, let's live in the country." [470]

Bill had arrived at Camp Tyson on an early and cold morning. "We arrived on the train. We'd been traveling for three days and three nights. It was cold, cold, and we walked from the gate to our barracks, about two miles. We didn't get to bed until 4 a.m.," Bill said. "That was my first introduction to the Army. We had KP right away the next morning." [471]

Bill was very busy, learning how to operate the barrage balloons. He said, "Camp Tyson was accused of being like a boy scout camp" by soldiers from other camps. "We had flowers on our camp and it did look nice, but we did a lot of training, a lot of walking. We were busy all the time" [472], including maneuvers, which involved launching balloons off of barges on the Tennessee River. [473]

---

[468] Ibid.

[469] Ibid.

[470] Ibid.

[471] Bill Perkins to Susan Gordon of the Tennessee Historical Society for "Home Front" project, April 17, 1992.

[472] Ibid.

[473] Ibid.

After his training at Camp Tyson was over, Bill was transferred overseas for a time and when the war was over, he opened an electric shop, Perkins Electric Service in Paris. He died June 23, 2002. The couple had two children, William, Jr. and James David.

Bill said in 1992, "I liked being here. After Camp Tyson, Paris had no way to go but up." [474]

"I SAID I MAY BE CRAZY, BUT I'M NICE CRAZY"

First Lt. Cal Orris saw Maxine Etheridge walking down the street while he was driving through Paris with some of his men.

"I said, 'I'm going to marry her. I didn't know who she was, but I was going to find out.'" And find out he did. "I called her that night and told her my plans. I told her I was going to marry her. She said, 'You're crazy.' I said, 'I may be crazy, but I'm nice crazy.'" [475]

It was a hard sell at first, but Cal was persistent. "I told her I was going to keep calling her until she agreed to go out with me. After two weeks, she agreed to go out on a double date with me. And we were married Sept. 30, 1942." [476]

When Cal died in 2001, the couple had been married for 59 years, building a life in Paris that included raising two girls, Susan and Jeri.

Susan Orris Palmer said her father was raised in Pittsburgh "but really loved the South. After my parents got married, they went back to his home in Pittsburgh but left after three months. Neither one of them liked it: he even didn't like it there anymore." [477]

Orris, who was an engineer with the 861st AAA AW Battalion, became a plant foreman at the Holley Carburetor Plant in Paris. At Camp Tyson, he had been a bass player in Tom Lonardo's camp band and he continued playing with Lonardo after the war. He joined the National Guard, retiring as a major.

"He just liked it in Paris," Palmer said. "He fit right in and immediately became involved with the community." [478]

---

[474] Ibid.

[475] Cal Orris to Susan Gordon, April 17, 1992.

[476] Ibid.

[477] Susan Orris Palmer, personal interview with author, March 11, 2010.

[478] Ibid.

LOVE BLOSSOMS OVER LAUNDRY

Anne Thompson worked in the office at the laundry at Camp Tyson. And as luck would have it, James Wilson was in charge of taking care of all the supplies for his company, and as such he would take bags of dirty uniforms to the laundry each week.

Promoted up the ranks, Wilson eventually became Staff Supply Sgt., then 1st Sgt. , for the medical detachment. "The boys in our unit, the orderlies, the cooks, wore white clothes and it was very important for the clothes to be white, so going to the laundry was part of my duties," he said. "And that's where I saw her." [479]

Anne Thompson, whose father was a county judge, got a job at the camp. "My father really didn't want me to work there, but I needed a job!," she laughed. "I took care of the accounts for the laundry. The soldiers would bring their laundry in to be washed and I kept account of how much each soldier owed." [480]

She recalled, "One day he came in and told me he wanted to see me. I remember our first date was seeing 'Gone With The Wind' at a theater (in Paris), and then we went out after to one of the soda fountains, for a coke or something." [481]

The couple dated regularly after that, marrying in 1942. The newlyweds lived off base, at the Herman Cravens home on Poplar Street. Wilson stayed at the camp until 1944, but his service to his country was not over. From there, he was shipped overseas, first to New Guinea, then to Luzon in the Philippines, then on to Okinawa, Japan.

While he was overseas, Anne took her meals at the home of Mrs. Mary Freeman near First Baptist Church in Paris. "A whole bunch of soldiers and their wives ate breakfast and dinner there," Anne said. "I remember she had a long dining room table and eight of us ate there. She kept pretty busy and it was nice there." [482]

Discharged in 1945, Wilson returned to Paris and was encouraged by his father-in-law, Judge Jim Thompson, to accept a position at the Kentucky-Tennessee Clay Co., which became his lifelong career. He became manager of the Tennessee division of the company and worked

---

[479] Shannon McFarlin, "Camp Tyson Brought James Wilson To Paris," *The Paris Post-Intelligencer*, December 14, 2004, page 5A.

[480] Anne Wilson, personal interview with author, April 16, 2009.

[481] Ibid.

[482] Ibid.

there 38 years. He and Anne had a son, James, Jr. and have been active in the community and First Christian Church.

Wilson said his experience at Camp Tyson was a milestone in his life in more ways than one. "I served my country and it brought me here to Paris." [483]

"HE MUST HAVE BEEN THE ONE SHE PICKED FOR ME"

Kathleen King was the match-maker of McKenzie, a small town a few miles down the road from Camp Tyson.

Even though McKenzie is in Carroll Co., it was still close enough for its citizens to share the excitement of the flood of soldiers to the local area. Especially Mrs. King, whose husband, Glen was the mayor.

The Kings owned an elegant home on Magnolia Street, which they called Magnolia House, and which was perfect for entertaining. The couple wanted to open the doors of McKenzie to the Camp Tyson soldiers, just as the doors of Paris had opened to them, so in the days before an organized USO was in operation there, they decided to use Magnolia House to host house parties for the soldiers.

And for Mrs. King, it provided an opportunity to play cupid.

One of her main targets was Maxine Bersey, who was "almost family" since she worked as a secretary at the U-Tote-Em store, which was owned by John King, the mayor's brother. "Every Sunday Aunt Cass would call me and say, 'I'm going to have a party and I want you to come.' But my mother didn't want me to date soldiers. I didn't want to tell her this, but I also didn't have a long dress and in those days you didn't go out without a long dress." [484]

Finally, Bersey said, she admitted to Mrs. King that the lack of a long dress was holding her back. "She said, 'Do you have a blouse?' I said yes. She said she would find me a skirt and call me later. [485]

Mrs. King did indeed find her a long taffeta skirt to wear and Bersey decided she needed a fancier blouse, so went to town and purchased a sheer white, long sleeved blouse. "Then I went to Aunt Cass' house that night and the first person I met was Kenneth Bersey. He sat close me, so he must have been the one that she picked for me." [486]

---

483 McFarlin, "Camp Tyson Brought James Wilson To Paris."
484 Maxine Bersey, personal interview with author, January 27, 2010.
485 Ibid.
486 Ibid.

Her matchmaking kept Mrs. King busy in those days since "the soldiers were always looking for girls. Southern girls must be different from Northern girls because they said they really liked the Southern ones because we were more shy." [487]

Maxine said she and Kenneth did not hit it off right away, but he persisted in asking her for dates. "I decided to ask him for Sunday dinner and I asked Momma if I could invite him. She said no, but I reminded her my brothers were in the service and she would want someone to be nice to them, so she finally said yes," Bersey said. The dinner went smoothly and Kenneth got high marks when he wrote a thank-you note to her mother, which included the request that he be allowed to see Maxine again. "So after that, he came every Sunday for dinner," she said. [488]

The two became "regular steadies," going to the picture show and to the service club at the camp. So much so that Bersey's mother came to check her out all the way from New Jersey. "I'll never forget my first sight of his mother, getting off the train at the L&N Depot downtown. She had on the most outlandish hat I'd ever seen. Even Kenneth said, 'Mother, where did you get that hat?'" Kenneth's mother said that her husband had sent her to check on things "because Kenneth could get with a girl down there and be ruined for life," she said. "But she must have liked me pretty well, as it turned out." [489]

The couple married in March, 1944, and for 22 years lived in New Jersey before returning to McKenzie. Living in New Jersey was a bit of culture shock for Maxine. "I never could get warm there and if we went to the store, people couldn't understand a word of what I was saying. It embarrassed me so." The couple returned to McKenzie. "He loved it in New Jersey, but he couldn't wait to move back here. He always loved the South and Tennessee. He liked the people, the weather, everything." [490]

Kenneth was a salesman before retiring. He died in 1986 at the age of 71, but their love story lives on in their daughter. "Ken always said if we had a daughter he wanted to name it after Mrs. King, who introduced us," Maxine said. [491] Thus, their daughter Kathy is a living reminder of the lady who sparked their romance.

---

[487] Ibid.

[488] Ibid.

[489] Ibid.

[490] Ibid.

[491] Ibid.

"DON'T TRY TO TRANSPLANT ME, I WON'T BE A YANKEE"

Like most young girls in Paris, the late Claire Taylor was part of the social whirl which surrounded Camp Tyson. The mothers of the girls of Paris formed a group called The Liberty Belle Club and they would charter Frank Blake's buses to take them to the camp for dances.

Wearing formal gowns, the girls would dance with the soldiers at one of the service clubs at the camp.

The late Tom Lonardo was the dashing bandleader of the Camp Tyson band and one night he saw something which gave him pause: Claire Taylor in her fancy gown. "He put his bass down and came off the stage. He told me right then he was going to marry me," Claire said. [492]

The couple did indeed marry, launching a partnership that lasted for decades. Lonardo became synonymous with music in Paris, operating a local music store for years and continuing to be a bandleader for his own dance band.

Lonardo's store became a hang-out for musicians and music-lovers, including Willie Nelson and Hank Williams, Jr. over the years. "Dad opened his doors to them and he'd say, 'Let's have fun.,'" his son, Tom, Jr. said. "He'd have a lot of musicians in on Saturday nights and they'd just play." [493]

Lonardo was from Providence, Rhode Island, and raised in a household where Italian was still spoken by his parents. Their son, Tom, Jr. of Memphis, recalled, "Mother went to Rhode Island to visit and everyone was speaking Italian and she didn't understand it. She didn't care for that too much. But my father fell in love with the South and that's where he wanted to live." [494]

Or as Claire herself put it, "I told Tom, 'Don't transplant me. I won't be a Yankee." [495]

Tom Lonardo said he eventually even learned to say y'all in his years in the South. "Paris was very good to me. It accepted me and I was able to have a good career here. My dance band played gigs in a

---

[492] Claire Lonardo to Susan Gordon April 17, 1992.
[493] Tom Lonardo Jr., telephone interview with author, April 16, 2010.
[494] Tom Lonardo, Jr interview.
[495] Claire Lonardo to Susan Gordon.

200-mile radius from here: even the Governor's Ball. But the most important thing is that Paris gave me my pretty wife." [496]

FINDING ROMANCE ON THE SIDE OF THE ROAD

A native of Connecticut, Howard Koenen never would have thought he would live his life in Murray, Kentucky, because of a war.

But that's what happened when he helped his future bride and her girlfriends get their car up and running one Sunday afternoon in Paris.

Koenen was a member of the 302nd Barrage Balloon Battalion and he and his buddies had driven into Paris from Camp Tyson. "We were on our way back, near Grove School, when we saw a car in the ditch and some girls looking under the car. And that's where I met my wife," he laughed. [497]

"We got some baling wire and fixed a rod for them and then followed them into Paris," Koenen said. "In those days the soda fountains were always open in Paris on Sunday afternoons, so we went to a soda fountain and one of the girls called her mother." [498]

The girls were from Murray, which is 22 miles from Paris, and the parents came to pick up the girls to take them back home. "The outcome was that the mother invited us to their house for dinner next weekend to thank us. She said, 'What do you like, chicken or steak? And my future bride was there, too, and that's how I got to know her.'"

The couple was married in November of 1942, and found an apartment off-base in Paris. "We found us a room in someone's house. There was one thing funny, the landlord was also renting a tree in his back yard," Koenen laughed. "I guess the law was you could only rent so many rooms in your house, so he also rented out his tree." [499]

After the war, the couple lived for a time in Connecticut, but his wife wanted to return to Murray. Koenen became active in civic life in Murray, serving on city council and various other commissions. "On my 90th birthday, I quit everything," he said.

Their marriage produced two daughters and lasted until his wife died in 2002.

---

[496] Tom Lonardo, Sr. to Susan Gordon.
[497] Howard Koenen, personal interview with author, October 30, 2009.
[498] Ibid.
[499] Ibid.

"He Just Came Up to Me and Started Talking to Me"

Oscar Gardner was a handsome man, especially so in his uniform and like many soldiers, he looked to the PX for entertainment.

That is where he met the pretty Mary Will, who already was enjoying a successful career as a teacher at the Henry County Training School.

"They were having a dance at the PX and he just came up and started talking to me," Mary laughed. "He asked me for my telephone number and I guess I must have given it to him." [500]

Mary said, "That must have been 1942 because we got married in 1943." Mary's family were caretakers at Maplewood Cemetery in Paris and they lived in a house on the cemetery grounds. "When he'd come to visit when we were dating, the Camp Tyson bus would drop him off at the road and he'd have to walk that big hill up to our house," she said. [501]

At Camp Tyson, Gardner served as the mess staff sergeant and after his service at Camp Tyson, served in the signal corps. Once married, the couple wanted their own home, but lumber was scarce during war times. "So we just moved that old house to a new lot." [502]

For Mary, the couple's early married life was filled with anxiety after her husband was transferred overseas. In an interview in 2004, Oscar said, "I was a supply staff sergeant and we followed the invasions, going from France, Germany and the Philippines — that was the worst — and then on to Japan, right after the bombs were dropped there."

Both Gardners recalled the anxieties of their separation. "I was worried about him. I would get some letters from him, not many, because in war times he couldn't write all the time." Oscar remembered it was a worry-filled time for him, too. "There were long, long, long nights. I went for days on those ships and all I could see was water."

Once Oscar, who was from Waverly, Tennessee, returned home, the couple settled full-time in Paris, where Mary continued teaching and Oscar eventually found a good job at the Tennessee Valley Authority (TVA), where he worked for 28 years before retiring in 1977.

---

[500] Shannon McFarlin, "Mary Will Gardner Taught Many of Paris Youngsters In 43-Year Career," *The Paris Post-Intelligencer*, May 4, 2004

[501] Ibid.

[502] Ibid.

The couple built a life full of activity, especially surrounding their church, Quinn Chapel AME.

"When I Saw Her, That Was Just It"

The black section of Paris, according to Aaron Dobbs, Jr. "was reputed to be a place with the best-looking ladies, in other words, a bunch of hotties." His mother, Joy Randle, was one of those good-looking women, he said. [503]

On weekends and in the evenings, soldiers from Camp Tyson — and even as far away as Fort Campbell in Clarksville — would get a car and share the cost of gas, coming to the "Black Bottom" area of Paris where there were several nightclubs and restaurants.

"My parents used to tell me that on weekends, you couldn't hardly walk down the streets for all the soldiers in town," he said.

Joy was working in a sandwich shop owned by Ernest Dumas on Washington Street "frying hamburgers and working the grill. My father stopped by there one day and he told me when he saw her, that was just it," Dobbs said.

Aaron Dobbs Sr. "tried to make conversation with her — along with other guys — but she just rebuffed his first attempts." Joy was raised by her Aunt Judy Belle, who finally said to her, "Shoot, Joy. You got a job. Marry him."

The couple married in Blytheville, Arkansas, Aaron Sr.'s home state, and Joy went with him when he was transferred to Japan after the war was over. While at Camp Tyson, Dobbs was a member of the Headquarters Battery, 319th, and was promoted to T-Sgt., in April of 1943. [504]

Aaron Jr. was born in 1949, while the couple was in Japan. "I guess that makes me Japanese," Dobbs laughed. "Well, nah. I did have Japanese citizenship, but I relinquished it when I was 18." [505] Dobbs' father was a great influence on him and he also went into the military, making it his career.

---

[503] Aaron Dobbs, Jr., telephone interview with author, February 24, 2010.

[504] "Promotions," *The Gas Bag*, April 28, 1943, page 12.

[505] Aaron Dobbs, Jr. interview.

Christine Reynolds and Val Umbach got to know each other because a friend of hers was dating a friend of his.

Finally, a friend of theirs told them, "Y'all like each other, so why don't you date? So we did." [506]

Christine was also dating a First Lieutenant from Camp Tyson and Val was a Private First Class. "I couldn't decide between them, but then all of a sudden, I liked Val better." Val later was promoted to 2nd Lieutenant at the camp, working with the quartermaster unit. [507]

Umbach was from Chicago and was stationed to Camp Tyson in one of the earlier units. "He was one of the groups that organized the camp," she said.

A colonel at the camp advised Christine not to marry a soldier "but I did and it lasted 21 years," she said. Val's family "couldn't understand me," she said. "They'd try to get me talking because of my Southern accent, but we always got along." [508]

After Camp Tyson, Val was transferred to Fort Knox and then Fort Benning, and Christine went with him. Then he was sent to serve in Europe for 13 months. "So I came back home to Paris to live with my parents, with the bulldog he had given me as a gift." [509]

Once the war was over, the couple lived for a year in Oklahoma, where Val had inherited land. Returning to Paris, Val became more and more interested in photography, which eventually became his life's work. "At first he was in the sporting goods business with my brother, then Val sold his interest and opened a photography studio," she said. [510]

In Paris, Val Umbach's name was synonymous with photography. Most everyone in town has old photographs in their home that are stamped "Val Umbach Photography" on the back. His studio on N. Poplar St. sold cameras, supplies and equipment and "did real well really quickly," she said. And he took photos of everything, from weddings to portraits. Christine noted that Val also was very musical "and

---

[506] Christine Umbach, personal interview with author, July 15, 2009.
[507] Ibid.
[508] Ibid.
[509] Ibid.
[510] Ibid.

he would often sing at a wedding and then jump down and take the photos of the wedding, too." [511]

The Umbachs had two children, Christopher "Kit" and Valerie.

A CAMP WEDDING

In wartime, many soldiers who wanted to marry could not get permission to go back home for their weddings.

In that case, improvisation was key.

That is what happened when Soldier George Frances Batchelder of Lisbon Falls, Maine, wanted to marry his sweetheart, Mary Anastasia Sylvester of Portland, Maine. A ceremony right on the Camp was in order for them.

The couple was married Sept. 5, 1942, at the chapel on the camp by the Catholic Chaplain Captain Rivers. According to their daughter, Ann Batchelder Gormon, the ceremony included a 5 p.m. mass with music by the camp organist. Best man was Lt. William Moulder of Paris and matron of honor was Helen O'Toole of Portland, Maine. The wedding dinner for the party of six was held at The Greystone Hotel in Paris and the couple honeymooned at The Peabody in Memphis.

Batchelder was in the 455th AAA (Anti Aircraft Artillery) and trained with the barrage balloons for three months at Camp Tyson. He first went to Camp Davis to Officers' Candidate School, was promoted to 2nd Lieutenant. He was promoted to 1st Lieutenant. in February, 1943, and left for Europe in Oct. 1943. He remained with the 455th until the end of the war.

While Batchelder was stationed at Camp Tyson, his wife took an apartment at the Barton Apartments in Paris. His daughter reflects, "Many of the sweethearts and wives followed their boyfriends or husbands wherever they were stationed in the U.S., knowing they would probably soon be sent overseas. The couples wanted to spend as much time together as they could knowing the future was very uncertain." [512]

BUILDING A LEGACY OF EDUCATION

It was a lucky day for the black community of Paris when Joseph Harden was assigned to Camp Tyson. A Rome, Georgia, native, he met Elizabeth Diggs of Paris, while he was stationed there and the two

---

[511] Ibid.

[512] Letter from Ann Batchelder Gorman to author, April 6, 2009.

became central figures in the Henry County Training School which educated generations of young people.

Elizabeth Chambliss of Decatur, Illinois, said she isn't sure when or where her parents met. "I assume it was at the USO,'" she said. [513]

With his GI bill, Harden finished his college degrees at Tennessee State University after the war and first was hired as eighth-grade teacher and assistant principal at the Training School in Paris in September of 1947. He was named acting principal in 1948, serving as principal from 1949 until the school closed in 1968. His wife taught elementary school there for many years. [514]

When the Paris Schools became integrated, Harden was appointed assistant superintendent and served as supervisor of guidance and attendance for the Paris Special Schools district. He retained that post until 1985. He was an unprecedented two-term president of Paris City Teachers Association from 1970 to 1972. [515]

At the CME Church in Paris, he served as Sunday School superintendent and he was manager and narrator of The Golden Voices, a local singing group, from 1963 to 1994, and he was heard with them for 25 years each Sunday morning on Paris radio station WTPR. [516]

Harden's impact was felt in the civic arena, as well, as he was instrumental in equipping Johnson Park in Paris with floodlights and organizing the first Little League team to play there. He was named recreational director of the park in 1952 and continued with that until 1978. [517]

"They really made an impact on the black community in Paris," Chambliss said. [518] Both Hardens are now deceased.

"I THINK ABOUT HIM ALL THE TIME"

The late Homer Sanders, Sr. was another man who never expected to live his life in Paris, Tenn. From Longneck, Texas, he was stationed at Camp Tyson and met a girl named Minnie Mae Stokes and decided to stay after the war.

---

[513] Elizabeth Chambliss, telephone interview with author, June 11, 2009.
[514] Obituaries, Joseph A. Harden, *Herald and Review*, Decatur, Illinois, April 16, 1999.
[515] Ibid.
[516] Ibid.
[517] Ibid.
[518] Ibid.

Elaine Sanders said her father "was a man of God. He taught us about Sunday School," but he didn't talk much about his service in the war. "He was very sweet and I think about him all the time." [519]

Both her parents are deceased now, but Sanders said she remembers her mother telling her she thought her father "was a good-looking man" and they started dating.

Homer — who his friends called "Red" — settled in Paris, working for a time at Camp Tyson after the war and later at the Holly Carburetor Plant. The couple raised four children.

SETTLING IN PARIS

Henrietta Holmes' father, Jacob Earl Nobels, met his future wife while she was working at Camp Tyson. "A lot of people in Paris had jobs out there and that's where they met," Holmes said. [520]

After his stint at the Camp was over, he was sent overseas, but returned to Paris after the war. "For a time, Daddy and Mother went to live in North Carolina, where he was from, but they came back to Paris," where they raised three children. [521]

---

[519] Elaine Sanders, personal interview with author, May 14, 2009.

[520] Henrietta Holmes, personal interview with author, April 14, 2009.

[521] Ibid.

Chapter 21

Hospitality, Southern Style

WAY from home and anxious about what more the war would have in store for them, soldiers at Camp Tyson could be a lonely bunch, especially on weekends.

Knowing that, Henry Countians in both the black and white communities opened their arms to them, inviting them to church services and Sunday dinners.

Jessie and Minnie Olive of Cottage Grove were one of the many local households that showed the soldiers the true meaning of Southern hospitality while they were stationed at the camp.

"Those young boys were so homesick and my mother was a good cook," J.J. Olive said. "So we would invite the soldiers over for Sunday dinner after church. That was a real big deal when the soldiers would come to our place." [522]

So precious are those memories that Olive still has photographs of the soldiers with his parents and other family members, as well as the heartfelt letters of thanks that the soldiers wrote to them.

The letters are especially interesting, since the Olives kept them in their original envelopes, with "Soldier, Sailor & Marine Center, Paris, Tenn." embossed on the return address.

Pvt. Victor Scozzard, of the Training Battery #3, wrote to the Olives on May 28, 1942. Enclosing photographs that the soldiers took during their visit, Scozzard wrote: "Well, folks we are giving you both a token of appreciation for your troubles and especially the splendid manner in which you showed all us boys the times of our lives. We honestly can't express in superlatives how happy we are for the grand dinner...It certainly makes us soldier boys happy and gay." [523]

Across the letterhead on the envelope, Scozzard wrote, "V for victory. We're in it. Let's win it." He asked the Olives to keep in touch and drop him a card of their own. The Olives did send cards to the soldiers and in another letter, dated June 1, 1942, Pvt. Sal Braccini wrote: "Received your lovely card today and we were more than

---

[522] J.J. Olive, personal interview by author, May 7, 2009.
[523] Pvt. Victor Scozzaro, letter to Olive family May 28, 1942, property of J.J. Olive.

pleased to hear from you. We were very happy to hear that you liked the small gift we sent and hope that we could have done better." [524]

Allowing the war to creep back into his thoughts, Braccini told the Olives, "We have this week and next at Camp Tyson and might make a move as soon as our schooling is over. We intend to be graduated next week and after that, who knows." [525]

Olive said the schoolchildren in the Cottage Grove area — several miles from Camp Tyson — were so excited about the soldiers "half of the Cottage Grove students would play hookey to go out to Camp Tyson and watch the soldiers." Driving into Paris was equally exciting during that time period, he said, "because Paris was just covered up with soldiers." [526]

The late Jim "Spider" Dumas, who was a long-time columnist for *The Paris Post-Intelligencer*, recalled that his grandparents also hosted soldiers at their Cottage Grove home. "I remember on more than one occasion accompanying my parents to my grandparents' house in the 1940s and seeing shy soldiers sitting in the living room or seated around the table for dinner." [527]

Most of the soldiers, Dumas recalled, "came from far-away states that we boys couldn't spell or pronounce, like Massachusetts and Pennsylvania. I recall an old friend telling me not long ago that he met his first Yankee eating dinner at his mother's home, and had trouble communicating."

SINGING FOR SUPPER

Not too many miles down the road, the households of Barr's Chapel African Methodist Episcopal (AME) Church were doing the same thing.

Dorothy Cook of Paris has photos of the soldiers who visited the church on Sundays for sing-alongs and Sunday dinners. A group of Camp Tyson soldiers were very good singers and would entertain the crowd. "It was a regular Sunday thing at Barr's Chapel," Cook said.

---

[524] Pvt. Sal Braccini, letter to Jessie and Minnie Olive, property of J.J. Olive.

[525] Ibid.

[526] Olive interview.

[527] Jim "Spider" Dumas, "Henry Countians Have Always Helped Others, Even Lonely Soldiers," *The Paris Post-Intelligencer*, July 7, 1995.

"But not just Barr's Chapel, the black soldiers would come to the different black churches all over the county." [528]

The soldiers could also enjoy musical entertainment presented for them by the young ladies of the local "Double V Club." The "Double V" was a national African-American organization, with the Double V standing for "Victory At War, Victory At Home." It was organized to promote integration and civil rights both in the armed forces and in the homeland.

Cook said, "The Double V Club would help out the soldiers by organizing the dinners for them." [529] The girls in the Double V Club would wear red, white and blue outfits and sang a mixture of gospel and other songs at the Sunday get-togethers. Cook has photographs that show the Camp Tyson soldiers' singing group as well as the girls in the Double V Club, along with soldiers standing with them. They are dated 1943.

One of the girls in the photos is Lena Bell Taylor, who recalled that the soldiers "had good voices. They were very polite and nice. Every one of them was nice." [530]

She said her sister, Rebecca Taylor, was President of the Double V Club, which would meet at each other's houses on a regular basis for meetings and to practice singing. "I'll never forget Lillie Hurt Lancaster. She was older than us and she's the one that wrote the rules we had to follow." [531]

Membership dues were 25 cents each and she said their meetings were rather routine. "We'd pay our dues and talk about what programs we were going to do for church. It mostly just gave us something to do," she laughed. "And we'd help sew our skirts that we wore when we sang. They were striped red, white and blue, and we wore them with white shirts. Usually when the soldiers came to church, the food would be spread out on tables and everyone would eat outside, she said. "Mamma always made dinner and she fixed everything under the earth. Corn, greens, cakes, chicken, peas," she said. [532] Often, she said, her family would invite the soldiers to their house for dinner, too.

[528] Dorothy Cook, personal interview with author, March 26, 2010.
[529] Ibid.
[530] Lena Bell Taylor, telephone interview with author, March 29, 2010.
[531] Ibid.
[532] Ibid.

Those Sunday get-togethers were a high point of her young life, Taylor said. "We were so proud of how we looked and we just loved having those soldiers come for those visits."

As If They Were Ours

Chapter 22

Haul 'Er Down

PAUL Soik may not be a household name, but it should be, as he was one of the most accomplished men to pass through the gates of Camp Tyson.

Soik, of Lyndhurst, New Jersey, used his considerable skills as a draftsman and artist to craft a career as one of the most called-upon illustrators of books after his war service. At his death in 1999, Soik had 165 paperback covers to his credit. He was renowned for the stark and striking quality of his illustrations, which graced the covers of magazines, dime novels and Harlequin Romances. Since his death, his cover art has been included in exhibits in Toronto and New York City.

While at Camp Tyson, Soik could not contain his artistic nature and frequently painted and sketched. One of his large paintings, "Haul 'Er Down" which depicts Camp Tyson soldiers training with the barrage balloons, "was somehow saved during the demolition of Camp Tyson after the war," according to Brig. Gen. Roland Parkhill of Paris. [533]

The painting was kept for 45 years by the owners and employees of the Spinks Clay Co., which purchased the Camp Tyson grounds after the war. It finally came into the hands of Parkhill and for several years was displayed at the Vernon McGarity Armory in Paris. It now is in the permanent collection of the Paris-Henry Co. Heritage Center in Paris.

"Haul 'Er Down," an oil painting on cotton measuring 45 x 40," was restored in the 1990s by Gene Bechtel of Trappe, Pa., whose clients included the National Gallery and the Metropolitan Museum of Art. The restoration was facilitated by Paris Artist Joe Routon Jr., who now lives in Haddonfield, New Jersey, and who carried the artwork to Bechtel.

---

[533] Roland Parkhill, personal interview with author, July 9, 2009.

Another of Soik's paintings during his Camp Tyson service, a painting of a Coast Artillery Corps battery in action, is included in the book, *Art in the Armed Forces*, published by Hyperion Press. [534]

For many years, Soik believed the painting was lost until he was contacted by Parkhill in 1997. Using Soik's signature on the painting and after an internet search, Parkhill was able to track Soik down at his New Jersey home. Thus began a correspondence between Parkhill and the artist until Soik's death. With information on his life after Camp Tyson provided by Soik in his letters, Parkhill dedicated the October 1997/March 1998 issue of "The Color Bearer" to Soik. Parkhill edited "The Color Bearer," a newsletter of the 5th Tennessee Volunteer Infantry Regiment Association.

Soik completed Art School on June 8, 1941, and left the next day for Fort Dix, N.J. to begin four-and-a-half years of service in the U.S. Army. After basic training, he was assigned to the Barrage Balloon Board as a draftsman and was first assigned to Camp Davis, N.C., and then on to Camp Tyson. [535]

When the board was disbanded, he was first assigned to a photographic outfit and then to the 598th Engineer Topographic Battalion, where he made schematic drawings. In his spare time at Camp Tyson and during his later service, he painted and sketched his fellow soldiers to send to their loved ones at home and he received commissions for portraits from officers to be placed in Service Clubs. [536]

In a letter to Parkhill dated August 29, 1997, Soik said after Camp Tyson, he eventually joined the 11th Airborne Division as a paratrooper, serving in the Philippines. "We were the first troops to enter Japan on V-J Day," Soik wrote, and he did numerous pencil sketchings of the aftermath of the Atomic bomb drop while in Japan. [537]

In a large drawing, which Soik called a 'personal history chart' of his wartime activities, Soik wrote that "V-J Day was overcast as a formation of fighter planes few overhead. On the horizon a flotilla of gray hulls were visible, one of which was the battleship Missouri," on which the treaty was signed. [538]

---

[534] Aimee Crane, *Art in the Armed Forces* (Hyperion Press, 1944).

[535] Karen Hughes, "Local Artist/Soldier Captured War On Canvas," *The Leader* (Rutherford, New Jersey) November 5, 1998, page 7.

[536] Ibid.

[537] *The Leader*, Nov. 5, 1998.

[538] Soik "personal history chart," property of Roland Parkhill.

On their first night in Japan, Soik wrote, "We slept on the corked floor of the Nippon Race Track and we woke up full of fleas the next morning." [539]

In another letter to Parkhill dated Nov. 12, 1997, Soik noted that while aboard the M.S. Weltevreden, a Dutch troop ship, "I was put on special duty to sketch the personnel aboard ship and at the end of this service, I was given a lunch and a cot up on deck" — as well as a ribbon award from the commander to add to his collection of some 11 medals and decorations he was awarded during his war service. [540]

Upon his discharge from the Army in 1946, Soik was able to embark on his career, noting in a letter to Parkhill that making a success as a commercial artist "isn't an easy life...it takes more than skill to make a go of it in the art field." [541]

During his correspondence with Parkhill, Soik was ailing with diabetes. Parkhill received the sad news of Soik's death from his wife, Mary, in a letter she sent Jan. 23, 2000. She said her husband "had pleasant memories of Tennessee and spoke very highly of the people. He was very happy to hear that his painting of the Barrage Balloon has been restored and donated to the Heritage Center and to know that he'll always be remembered." [542]

---

[539] Ibid.
[540] Soik letter to Parkhill, Nov. 12, 1997.
[541] Ibid.
[542] Mary Soik, letter to Parkhill, Jan. 23, 2000.

Chapter 23

From Surplus To POW Camp

IN August of 1944, Camp Tyson was declared surplus by the War Department and plans were drawn up to prepare the facility for use as a large prisoner of war camp. [543] According to details gleaned from documents from the State Department, since April of that year Camp Tyson already had been housing a small contingent of prisoners, some 250 men.

A letter from Army Service Forces headquarters in Atlanta to General Maynard dated September 2, 1944, outlined the new plans for the camp. With the subject line, "Disposition of Camp Tyson," General Maynard was informed that Camp Tyson "has been declared surplus by the War Department, effective 14 August 1944, and all Service Command personnel, supplies, equipment, and activities will be evacuated therefrom at the earliest practicable date." [544]

Further, it was ordered that all real property at the camp, including its buildings and fixed installations, would be transferred to the U.S. District Engineer in Mobile, by October 15, 1944. Enlisted men would be transferred to the surplus detachment, 1457th SCU. [545]

Attached to the letter was a report which outlined plans for conversion of the camp to a "large Prisoner of War Camp." The letter acknowledges that the plans were made "in a short time and without benefit of a field survey. If the camp is to be seriously considered for housing Prisoners of War in large numbers, it will be necessary that a field survey be made by this office before a dependable layout can be prepared, before an exact estimate of cost can be made, and before the housing capacity which can be economically fenced can be determined." [546]

However, "based on this hasty survey," the letter continues, "it is believed that approximately 12,000 Prisoners of War can be housed at

---

[543] "Disposition of Camp" letter, NARA, Tyson, Camp, Tn. ARC Identifier 899615/MLR Number A1 461, RG389, Entry 461, Box 2510, Location:290/34/24/04.

[544] Ibid.

[545] Ibid.

[546] Ibid.

this station in addition to the existing 250 man Prisoner of War camp."[547]

PLANNING FOR PRISONERS

The War Department detailed what would be needed for the conversion:

For the Compound itself, it was estimated that fencing totaling $30,075 would be required, as well as 20 guard towers ($44,000); six buildings ($24,000); lighting ($46,000); roads and security bars ($6,000); and miscellaneous conversions ($5,000). For the Hospitals, it was estimated there would be a need for four guard towers ($8,800); and more fencing ($7,200). Adding in 15 percent for contingencies and 10 percent for overhead, the total estimated cost to convert the facility to a prisoner of war camp was $216,409. [548]

Attached to the letter was authorization from B.W. Bryan, Brigadier General, Assistant Provost Marshal General, for the conversion. "In order that the theaters, administration buildings, post exchanges, workshops, recreation buildings, service club and officers' quarters may be utilized to the maximum for prisoners of war, it is requested that the attached layout plan (enclosure two) be substituted for the tentative layout plan prepared in the Fourth Service Command." [549]

By the end of World War II, some 425,000 German, Italian and Japanese prisoners of war (POWs) were imprisoned in over 660 camps that were located in almost all of the 48 states that made up the United States at that time. [550]

In Tennessee, Camp Tyson was not the only facility housing POWs. Several other facilities, including Camp Shelby and Camp Forrest and a facility in Crossville, were utilized.

The Geneva Convention, as ratified in July of 1929, set forth the conditions under which prisoners of war could be incarcerated and regular inspections were made to POW camps in the United States to ensure that the conditions were met. [551]

---

[547] Ibid

[548] Ibid.

[549] Ibid.

[550] *Held in the Heartland: German POWs in the Midwest, 1943-46* (TRACES self-published n/d), introduction page.

[551] Ibid, page 4.

AS IF THEY WERE OURS

POW camps were required to provide adequate food, clothing, medical, sanitary services, provisions for religious, intellectual and physical activities, according to the terms of the Geneva Convention. Additionally, treatment of prisoners was to be afforded according to rank, with pay rates for POWs established for the work they would be performing in the camps. [552]

TYSON PRISONERS TREATED WELL

Inspection reports for the Camp Tyson POW camp are today stored in the Library of Congress and State Department and overall, reflect well on the treatment that its prisoners were afforded.

In an inspection conducted September 13, 1944, by Dr. Rudolph Fischer, representing the Legation of Switzerland for the Geneva Convention, and John Brown Mason of the State Department, the Camp Tyson POW camp was still housing 250 men. [553]

The inspectors reported there were 16 beds for POWs at the camp hospital, which saw an average of one patient per day. Recreational facilities consisted of one field for handball, "not large enough for football." [554]

Religious services were provided every Sunday in the recreation hall, with Catholic and Protestant American chaplains "who do not speak German." Despite that, the report said, "Attendance quite good." [555]

When the inspectors arrived, the POWs were being utilized "on post maintenance...connected with the closing of Camp Tyson. The Camp Commander (Capt. George Williams, Jr.) considers the work of the prisoners 'excellent.'" Average earnings for POWs during August, 1944, were reported as $23.80. Inspectors determined that "Prisoners now get the food they like (not the regular Army issue). All work details eat in the mess hall." [556]

No punishment of prisoners had been necessary since the POW camp opened in April, 1944, it was reported. The guard company had

---

[552] Ibid.

[553] State Department document, stamped "From The Special War Problems Division, Department of State" dated Nov. 27, 1944, property of Chris Corley.

[554] Ibid.

[555] Ibid.

[556] Ibid.

a strength of 48 men, including five administrative personnel. The Swiss representative told the Camp Commander "that he wished to make the 'best comments possible on the camp.' Further, the Swiss Representative reported "he considered it one of the very best small camps out of about one hundred that he has seen." [557]

By the end of August of 1944, hundreds more POWs had been transferred to Camp Tyson and "Prisoner of War Camp Labor Reports" begin to be filed with the U.S. Army. Several of the inspection reports are on file with the National Archives and Records Administration in Washington, D.C. and provide a glimpse of the type of work being performed by the POWs at the camp.

The reports indicate that the bulk of the prisoners held at Camp Tyson were noncommissioned officers and privates, with no officers housed there during that late summer and fall. A typical labor report, filed August 31, 1944, shows 288 noncommissioned officers and 2,896 were housed at the camp for that report, for a total of 3,184. [558]

Of that number 39 had been hospitalized and unable to work; 31 were on sick call; seven were non-working noncommissioned officers and 141 were listed as "essential unpaid labor." Of the 2,583 who were available for paid work, 2,564 of them worked four hours or more a day. By projects, the POW's work was categorized this way: 126 on company overhead; 42 worked in the canteen; 42 in maintenance and supply; 70 in the bakery; 197 on clothing and equipment; 847 on other post quartermaster division; 152 on buildings and utilities; 188 on grounds and roads; 502 on other post projects; 43 on motor maintenance; and 355 for EM mess. [559]

In remarks attached to the report, it was indicated that 96 POWs worked on KP duties, with 32 as bakers. Thirteen worked in the post laundry washing POW clothing. [560] According to all of the Camp Tyson labor reports, none of the local POWs were involved in private contract work outside of camp (that type of work would typically involve construction, logging, forestry and road work). [561]

---

[557] Ibid.
[558] Prisoner of War Camp Labor Report, dated August 31, 1944, NARA, ARC Identifier 899615/MLR Number A1 461, RG389, Entry 461, Box 2510, Location: 290/34/24/04.
[559] Ibid.
[560] Ibid.
[561] Ibid

However, John Yenchko did remember that the Camp Tyson POWs did help with farm work and drained the surrounding swampland. He said they were good workers, who were grateful for the good food at the camp, which was a vast improvement over that served in the German camps." [562]

As indicated in the labor report, the higher ranking officers were not required to work, although after the Nazi regime capitulated near the end of World War II, some began to volunteer for work so that they could be paid. [563]

DAYS REGIMENTED

According to Irvin Kellman, exhibit guide for the TRACES mobile exhibit which displays POW history and memorabilia, the day-to-day life for POWs in the United States was surprisingly similar.

POWs, Kellman said, were paid 10 cents an hour, up to 80 cents a day, and were "paid in camp script. No hard cash was given to them." Normally, he said, POWs would use the camp script in the camp canteen "to buy things that were rationed and that others might not be able to get, such as salt, sugar, flour, tobacco. They also used their camp script to buy their art supplies. The Red Cross provided them with instruments." [564]

As evidenced by the documents in the State Department and Library of Congress, Kellman said each camp "kept meticulous records...because they needed to uphold the Geneva Convention. Each camp meticulously recorded the name, rank and serial number of each prisoner." At most camps, POWs were expected to arise at 5 a.m., eat breakfast from 5:30-6 a.m., after which they would work until 5 p.m. Dinner periods would be from 5:30-6 p.m., he said, with free time from 6-9 p.m. [565]

For the most part, he said, POWs appeared to have been treated well in the United States since between 20,000-25,000 of the former prisoners returned to the United States to live after the war. [566]

---

[562] John Yenchko, interview with Susan Gordon of Tennessee Historical Society, March 23, 1992. Quoted in "Tennessee Historical Quarterly," published Spring 1992, page 13.

[563] *Held in the Heartland: German POWs in the Midwest*, page 18.

[564] Irvin Kellman, personal interview with author, Oct. 6, 2009.

[565] Ibid.

[566] Ibid.

PLANTING GARDENS, SINGING, AND CARVING

At Camp Tyson, POWs seemed to have made the most of their free time, with many of the German prisoners using their time to create art work, especially paintings, and for wood-working. The few Italian prisoners who were housed at Camp Tyson also spent their free time planting gardens in which they planted herbs such as oregano which were familiar to them. During a tour of the grounds at the former Camp, Spinks Clay Employee Chris Corley pointed to a ridge where wild oregano that was planted by the Italian prisoners still grows. [567]

Eddie Clericuzio was still a soldier at Camp Tyson when the Italian prisoners arrived and his knowledge of Italian was put to use as a translator for the prisoners. "We had a few Italian prisoners there and about 10 of them couldn't speak English I remember. So I was kind of put in charge of them." The prisoners, he said, "were never mistreated. They actually ate the same food we did." [568]

Howard Koenen was still stationed at Camp Tyson when the German prisoners arrived and he recalls that "Most of the German prisoners we had there were from the Afrika corps. They were involved in the brigade that was with the "Desert Rat.' I can remember they all had fair complexions. They were the cream of Hitler's Army and they were captured during the African campaign." [569]

The prisoners stayed to their own areas, Koenen said, "but I can remember hearing them singing at night. They were singing German songs and you could hear them all over the place." [570]

Red Boden of Paris can remember hearing Esther Graham speaking about the German prisoners' spending habits. Graham operated the former Sears store in Paris into which the prisoners would be allowed to come periodically.

"She told me that the German POWs were convinced the Germans would win the war so once a month they would go to Sears and use their script to order diamonds. They were sure the U.S. economy would go down and they didn't trust American money, so they con-

---

[567] Chris Corley, personal interview with author, May 22, 2009.
[568] Eddie Clericuzio, personal interview with author, May 11, 2009.
[569] Howard Koenen, personal interview with author, October 30, 2009.
[570] Ibid.

AS IF THEY WERE OURS

verted their script to diamonds which they thought would keep their value," Boden said. [571]

The late Ethel Murphy worked at Camp Tyson and she came into contact with the German prisoners periodically. She worked as a clerk typist at the camp dental lab and recalled that the POWs were sent to the lab a dozen at a time for dental check-ups.

She recalled the difficulty in recording the name, rank and serial number of each POW, which was camp policy. "We finally learned that one of the prisoners spoke English. That helped a lot," she said. [572]

Murphy said the first group of prisoners used to perform military salutes when they approached the U.S. Army officers in the dental clinic. "Finally, Col. Wycoff told the prisoners they were not required to salute." [573]

PRISONERS' PAINTINGS LIVE ON

Bennye Phillips of Hazel, Kentucky, has good memories associated with the German prisoners — and beautiful paintings to enhance her memories.

Phillips' father, Claude B. White, Sr., was hired as an overseer in the camp's last days and he became good friends with the officers still assigned to the POW camp. "One of the German prisoners there — his name was Wilhelm Sachs — had done an oil painting of the General's wife. My Dad saw it and asked, 'Do you suppose you could get him to paint another one of my wife?' Dad took a photograph of Mom and the prisoner did an oil portrait of her, too." [574]

The portrait still hangs on the wall at Phillips' home, as does another painting, of a German castle, which Sachs painted, too. [575]

Bobby Freeman of Paris remembers that when his brother worked at the camp, one of the prisoners made a wooden gun for him. [576]

As with any situation in which men were confined against their will, all was not rosy at all times.

---

[571] Red Boden, personal interview with author, November 12, 2009.

[572] Jim "Spider" Dumas, "Camp Tyson Days Recalled," *The Paris Post-Intelligencer*, April 16, 1992.

[573] Ibid.

[574] Bennye Phillips, personal interview with author, July 3, 2009.

[575] Ibid.

[576] Bobby Freeman, personal interview with author, April 22, 2009.

Elliott "Eddie" Moody, who was a civilian military policeman at Camp Tyson, remembers at least one occasion when prisoners tried to escape the camp. When three escaped and were returned to the camp three days later, Moody was stationed as a guard at the main gate.

"Those men were sure glad to get back to the POW camp," he said. "I think they must have wandered about in the woods in the Palestine area, living on nuts, berries or whatever they could scrounge." [577]

The POWs, Moody said, "could swim in that little lake out there, and they did used to march them around, but they didn't really do much work. They were treated super, not mistreated. Actually, I thought they were treated better than the soldiers." [578]

Officers who had been stationed at Camp Tyson and who later worked at POW camps elsewhere also did not have such positive experiences.

TRYING TO ESCAPE

The late Major Herston Cooper, author of a self-published book, Remembering Camp Tyson, was commander of the Crossville POW camp after his service at Camp Tyson was completed.

Cooper was sent to Crossville in 1943 to command a stockade containing about 1500 German and Italian prisoners. All were officers, except for a few enlisted prisoners. According to an interview with Cooper published in The Knoxville News-Sentinel in January of 1975, Cooper most likely was sent to Crossville because those previously in command had not been deemed firm enough with the prisoners. Cooper was selected for the Crossville job because of his background in police and prison work.

According to The News-Sentinel article, Cooper was confronted with former Italian generals and former Nazi officers who were demanding bigger daily beer allowances and valet services, among other favors. [579]

Cooper's refusals did not endear him to the prisoners, who took out their frustration by trying to tunnel out of the camp. When the prisoners were captured trying to escape, a physical fight ensued, during which Cooper was attacked by prisoners led by Major Eric von

---

[577] Dumas, The Paris Post-Intelligencer, April 16, 1992.

[578] Eddie Moody to Susan Gordon of The Tennessee Historical Society for "Home Front" project, April 17, 1992.

[579] Carson Brewer, The Knoxville News-Sentinel, n/d.

Graf. Von Graf and several of his fellow prisoners jumped on Cooper, chopping him on the neck and restraining him in a hammerlock. [580]

In the melee, Cooper kicked von Graf in the groin, cutting his intestines with his heavy boots. Later in the camp hospital, von Graf "screamed that he wanted to die for the 'Fuehrer' and died of peritonitis two days later. [581]

Cooper also was treated in the hospital for injuries to his arm and shoulder, but in later years determined that the chop to the back of his neck had broken a vertebra which had steadily worsened over time. [582]

Over the years, it has been rumored that several prisoners were buried at Camp Tyson and still remain there, in unmarked graves. A former graveyard still does exist on the grounds and many believe that is where the prisoners were buried.

Today, the old cemetery is still adorned with a brick walk around it and a walkway up to it, both of which are deteriorating. No gravestones still exist, but a couple of people, both of whom wished not to be identified, said they remember there were gravestones in the cemetery after the camp was closed.

A man who worked at the camp said he remembers several stones still at the cemetery in the 1960s. "I can remember several stones that were still there. They were marble with black lettering and they had numbers written on them. I just wondered if they were the prisoners' identification numbers? I don't think they were soldiers' graves because the soldiers would have been sent off to their families after they died." [583]

Dickie Carothers, whose family purchased the Camp Tyson property after the war, said no graves were found on the camp when his family bought it. [584]

The News-Sentinel interview with Cooper also seems to dispel that belief. Cooper related that most of the German POWs who died while incarcerated in other camps in the U.S. were buried in the Crossville facility. According to the article, some 80 German prisoners were buried there in all and all were "shipped back to Germany after the war." [585]

---

[580] Ibid.

[581] Ibid.

[582] Ibid.

[583] Anonymous source, personal interview with author.

[584] Dickie Carothers, personal interview with author, August 3, 2009.

[585] Carson Brewer, *The Knoxville News-Sentinel*.

Chapter 24

Didn't You Hear?
The War Is Over!

WORLD War II was a long, exhausting and frightening experience for all Americans. When the war was declared over, there was a shared knowledge of relief and unbridled joy that swept over the country, from every big city and into all the small towns and villages.

In Paris, car horns started honking and the train and fire whistles blew to herald the good news.

For Virgil Wall of Paris, that day was still a vivid memory many years later. "It was August, 1945, the day that they came to call V-J Day, and I was working at the L&N Railroad shop," Wall said. "I remember I was walking home after work and every whistle at the shop started blowing. I kept walking and I heard the fire whistle blowing downtown. Then I began hearing car horns honking. I knew something was going on, but didn't know what." [586]

Wall began walking faster toward home and waved a car down. "I asked them what was going on. He had his car radio on and he said, 'Didn't you hear? The war is over!' That's the first I knew," he said. [587]

From her home on Washington St., Emily Daniel Cox could hear "people screaming and crying. Car horns were honking. It was thrilling. The war was over." [588]

But Wall couldn't spend the day celebrating. "I hurried right home and when I got there, there was a message for me from Capt. Hartsock of the Tennessee State Guard. He was the commander in Paris and I was a member of the state guard." [589]

The message informed Wall to report, along with the other local members of the guard, to Paris city auditorium at 6 p.m. There was to be a "big celebration downtown," he said, and the guard had been called to provide security.

---

[586] Virgil Wall, personal interview with author, May 12, 2009.
[587] Ibid.
[588] Emily Daniel Cox, personal interview with author, September 10, 2009.
[589] Virgil Wall interview.

"I got my uniform on and left for the auditorium," Wall said. "There was a lot of hustle and bustle and excitement in Paris and the powers that be were concerned with the celebrants. They wanted added security that night." [590]

Guard members were issued rifles and were ordered to specific stations to guard. As luck would have it, Wall's station was right in the middle of the celebration, which he could witness and enjoy, even while providing security. "My station was the corner of Market and Washington Streets, right in front of J.C. Penney's," Wall said. "A pretty good spot for watching everything." [591]

Wall said West Washington Street was blocked off and "an orchestra was called and a street dance was pulled together very quickly." [592]

Jo Hurdle Wall, Virgil's wife, grew up on Washington St. and said, "Everybody got out that night. The town was cram full. There were dances and parties — things going on all over." [593]

Virgil said the orchestra was placed on a raised platform facing the Crete Opera House (presently the county courthouse annex). "There was music, dancing. Everyone was in a festive mood," he said. "I was glad to be on that spot. I got to see the most activity of all the guys on the guard," he said. "We had someone stationed on every corner all around the court square and beyond." [594]

The spontaneous festivities lasted well after midnight, he said. And, no, there was no rowdy behavior that night. Smiling at the memory, he said, "Everyone was well behaved. Nobody caused any trouble. Everyone was just too happy." [595]

[590] Ibid.

[591] Ibid.

[592] Ibid.

[593] Jo Wall, personal interview with author, May 12, 2009.

[594] Ibid.

[595] Ibid.

Chapter 25

"It Was Considered Sinful"

O N April 20, 1945, Camp Tyson received official notification it would be disbanded and all troops still being housed there would be relocated.

Even before that, however, plans already were in the works for the camp's closure. After the D-Day landings, barrage balloons were considered obsolete and Camp Tyson's usefulness in the war effort was coming to an end.

Late in the war, Camp Tyson became a staging area for soldiers who were being transferred overseas, a well as an area for wounded soldiers to convalesce. It also was still being used as a prisoner of war camp.

The decision was made to salvage some of the buildings on the camp for materials that were still needed for the war effort. The federal government began notifying property owners that they would have first chance to buy back their land.

It was a sad time. The scene of so much excitement and important activity was no more.

Decisions made by the government during this time were controversial and considered wasteful by many. Many of the hundreds of buildings on the site were sold for scrap or just bulldozed and buried, as were caches of food and inventory.

Jonathan Smith of Jackson, Tennessee, whose father, Herschel, was a contractor during the camp's construction, said, "As a child, I was literally horrified as I heard the adults lament and criticize the government at the end of the war as they bulldozed big pits and cast quartermaster goods, tools and equipment into them. Presumably, this would hinder black marketing and keep the price of such things higher than if they'd been left in greater supply. In a time when much of the world was in hunger, the throwing of food into pits to rot was considered 'sinful' to some people or simply a great waste to others." [596]

---

[596] Jonathan Smith, letter to author, Feb. 1, 2010.

"He Salvaged as Much as He Could"

Bennye Phillips of Hazel, Kentucky, can attest to the amount of goods that were thrown out, as many of the items are still in her possession.

Phillips' father, Claude Barber White, Sr., was hired as an overseer of the camp in its last days "and they were just dumping things out over there," she said. "My father really didn't like that and he salvaged as much as he could." [597]

She still has her father's payroll receipts, showing he was listed as 'foreman over maintenance of Camp" until March 15, 1946.

Included in the items she has are manuals governing chemical warfare, how it is to be used, and how chemicals are to be formulated, as well as more mundane payroll and property accounts.

"The officers who were still out there used to give Dad things, like toilet paper, to bring home. We used that Camp Tyson toilet paper for quite a while and in those days, you needed things like that," Phillips said. Her father also was given baseball gloves, ball bats and balls that otherwise would have been thrown away. [598]

Phillips proudly displays a wool cap with Camp Tyson insignia, towels, pots and pans, a first aide kit, knife, and a kit with small tools. Her father became good friends with Camp Commander General John Maynard in the last days of the camp's operation and the general also gave him items from his office, such as chairs.

German prisoners of war were still at the camp while White worked there and they befriended him as well, creating paintings for him of a castle and of his wife, which still hang at Phillips' home.

A Virtual Ghost Town

Once so full of life and the site of such promise for Paris and Henry County, Camp Tyson became a virtual ghost town.

In May of 1947, Louis Newman of *The Parisian* newspaper, took a photographer and a small plane over the site and described what he saw in a May 23 article. "A scene of almost utter stagnation," Newman wrote of the camp. [599]

---

[597] Bennye Phillips, personal interview with author, July 3, 2009.
[598] Phillips interview.
[599] Louis Newman, *The Parisian*, May 23, 1947.

As If They Were Ours

"A guard's jeep on a street near the big steel hangar was the only sign of life or movement. Piles of scrap and waste materials were heaped about. Hundreds and hundreds of bare foundation block — grown up in weeds — freckled the ground below. Several big circles could still be distinguished in the grass where barrage balloons once were moored," he wrote. [600]

The remaining buildings "and the big hangar and the water tower stood out giant-size upon the leveled landscape, as did the several clusters of temporary structures left standing, such as the warehouse area around the railroad spurs, the old personnel building, the fire house." [601]

Bricks and scrap were not the only materials left behind.

MERCURY RUNNING LOOSE

Phillip Steele, whose family bought back their land that had been acquired by the government for the camp, recalls, "When they left, they sealed a lot of the concrete doors out there. Makes you wonder what was in there. We always heard they had buried the helium gas out there." Behind the huge hangar, which still is on the site, "you could walk down a metal door and into an underground bunker," Steele said. [602]

"You also could get mercury right out of the ground that was left behind," Steele said. "It was just running loose, like jelly." Steele said he sold some of the mercury to a science lab. [603]

The possibility of toxic chemicals being buried in the ground was a source of consternation for the late Roy Goins and he spent quite a bit of time and energy investigating it. In the 1980s, he determined that much of the Camp Tyson records were stored by the General Services Administration in East Point, Ga.

Goins was concerned because he and his wife, Rebecca, owned land that her family had purchased back from the government, just a few feet from the camp's Gate 3 entrance. "He went through those records to see if chemicals were under the land. The records said there was nothing buried there," she said. [604]

---

[600] Ibid.

[601] Ibid.

[602] Phillip Steele, personal interview with author, May 18, 2009.

[603] Ibid.

[604] Rebecca Goins, personal interview with author, October 2, 2009.

Until her death in September 2010, Rebecca Goins had kept a suitcase full of legal documents that her husband accumulated during his protracted legal battle with the camp's present owner over right-of-way issues. Among them are an appraisal conducted by Dobson-Bainbridge Realty Co. on April 9, 1947, of the Camp Tyson property. Listed are several pages of inventory, along with the salvage value, of buildings on the Camp property. Included are such items as balloon hangar, replacement value, $110,000, salvage value, $7,474; repair shop, replacement value $45,600, salvage value $4,560; six gas storage sheds with replacement values of $4,800 each and salvage values of $480 each; pump house and pumps, replacement value of $3,587, and salvage value of $150. The appraisers determined that sold in five tracts, the land had a market value of $81,040. [605]

BALLOONS NOT FLYING ANYMORE

Maynard Cook recalls that his father, A.C. Cook, was one of the men hired to tear down the camp. "My father worked there for almost a year and the floor collapsed in one of the buildings they were tearing down. He fell down to the sub floor and a big iron piece, like a crowbar, was stuck in his side." His father "was laid up for a really long time because of that. He had bandages up on his arms and I can remember hearing him scream when Mom had to take the bandages off." [606]

In the months during which material from the camp was being salvaged, many Henry Countians purchased wood, iron — and even entire buildings — for their own properties. On Cook's property on Hwy. 54 sits a Quonset hut that once was used at the camp. It now is used for cattle on his farm.

"We lived out in the country, where we could see the balloons flying when the camp was operating," Cook said. "After Dad got injured, I can remember when I was a little boy, standing in the front yard and looking over toward the camp and wondering why I didn't see the balloons in the sky anymore. I didn't realize the camp had been shut down." [607]

---

[605] Appraisal of Camp Tyson property, April 9 1947, property of Rebecca Goins.

[606] Maynard Cook, telephone interview with author, March 30, 2010.

[607] Ibid.

Jerry Ridgeway of Paris, whose mother had worked at the camp, remembered its closing days with sadness. "It was a big letdown when they disassembled the camp. Everyone was glad the war was over, of course, but you missed the excitement." [608]

Employees of the camp were especially sad with its closure. Jeanne Townsend of Paris had worked as a secretary for several officers and said, "We were the last people to leave when the camp closed. Only the station complement was staying on and some of the girls and boys had a going-away party for us."

Townsend said, "I remember we put a black wreath on the gate at the camp as we left." [609]

---

[608] Jerry Ridgeway, personal interview with author, May 1, 2009.
[609] Jeanne Townsend, personal interview with author, June 2, 2009.

Chapter 26

"Don't Come Back From Washington Without A Signed Deed."

FOR a year after it closed, the Camp Tyson stood empty. No activity, piles of scrap were piled up, weeds grew unfettered.

Henry Countians were becoming restless. What was to become of the huge reservation that once showed such promise for the local economy?

The public saw an empty camp but behind the scenes, negotiations for the property were underway. The negotiations were protracted and full of intrigue. In the end, the highest bidder — the H.C. Spinks Clay Co. — won the bargaining, paying some $30,000 more than its opening bid, but some $81,000 less than the appraised value.

The War Assets Administration began advertising for bids "for sale or lease of surplus real property and facilities," setting a deadline of April 3, 1947, from federal government agencies or state and local governments; or until April 25, for all others. The bids were publicly opened at 10 a.m. CST, April 25, 1947. [610]

After selling 300 acres of the property back to its original owners, the government was offering 1,298.9 acres for sale on the "Camp Tyson Military Reservation." Included in the package were "water mains, sewers, incinerator, sewage disposal plant, railroad siding, adequate phone facilities, network of hard surfaced roads, electrical distribution system, deep wells, pumping stations and water treatment plant, approximately 500,000 gallon elevated steel water tank, steel balloon hangar with two HP line pressure boilers, stem heating system with radiators and blowers, and approximately one-quarter million square feet of warehousing space, together with other buildings and improvements." [611]

---

[610] War Assets Administration advertisement, "Invitation For Bids For Sale or Lease of Surplus Real Property and Facilities," property of Dick Carothers.

[611] Ibid.

Two main bidders emerged:

The Spinks Clay Co., which had been engaged in the mining and processing of ball clay since the turn of the century and had been operating a large-scale cattle farming business. The Spinks Co. already owned some 4,085 acres of land, located in the vicinity of Camp Tyson, as well as in Weakley and Carroll Co.

The Spinks Co. operated from Puryear, Tennessee, which is 10 miles from Paris, with an office in Cincinnati. Partners in the business were R.B. Carothers, Harriet (Spinks) Carothers, R.B. Carothers, Jr. and Harry S. Carothers.

The company already owned mineral and mining rights on 8,552 acres and were engaged in mining clay deposits in and around Puryear, Paris and Gleason, Tennessee. (in Weakley Co.) [612]

In its proposal, which was formulated Sept. 17, 1946, the Spinks Co. proposed to help produce business enterprises on the property which would include a wood-working plant to manufacture toilet seats, warehouse for farm equipment, manufacturing plant for common and facing brick, and a dairy and meat processing plant. These would be in addition to the clay mining and processing and cattle farming businesses the company was keenly interested in establishing there.

In its official bid, which was submitted April 23, 1947, Spinks pledged that with acquisition of the Camp Tyson property, its ball clay company could increase from a current 56 employees to 300.

Spinks initially offered $171,000 for the property and through the course of negotiations raised its bid to $181,000. The final price paid by Spinks for the property, as demanded by the War Assets Administration, was $201,000.

The Young Business Men's Club of Paris, which operated before a Chamber of Commerce became established for Paris. Key persons involved in negotiations for this organization were B.G. Diggs, F.N. Travis, and Brown Morris, president of the club.

The Club submitted its bid to the War Assets Administration on April 26, 1947, noting that "the conditions set out in advertising the sale of the Tyson property give preference to a community wanting buildings for employment for people. We think (our bid) clearly sets out this condition for Paris, Tenn." [613]

---

[612] "Proposal To Purchase Camp Tyson" submitted by H.C. Spinks Co., September 17, 1946, property of Dick Carothers.

[613] Young Business Men' Club proposal, April 26, 1947, property of Dickie Carothers.

The club stated it had "several prospective manufacturers, with whom we have advanced to various stages of negotiations regarding local plants at Paris," which included the American Lady Corset Co. of New York; the John Hary Leather Co. of St. Louis; the Burd Piston Ring Co. of Chicago; and M. Fine and Sons Mfg. Co. of New Albany, Indiana.[614]

The proposal went on, "Although our competitor wants all of this area, we do not think there is a doubt but that if we were awarded No. 1 and he were offered the balance he would accept as soon as he was convinced that No. 1 was not available for him, and we believe that should we be awarded No 1, it would greatly enhance our chance to secure some industrial development and also not interfere with our competitive bidder's usage of the remainder."

The club offered $37,000 for the property.

A STUMBLING BLOCK EMERGES

A major stumbling block in the sale was a request from the National Guard to remove Building #136 from the Camp Tyson reservation and move it to the University of Tennessee Junior College at Martin, Tenn. The National Guard's initial request was made in September of 1946, asking for three buildings to be moved to Paris, Martin and Union City, Tenn., but by February of 1947, the request had narrowed to one building to be moved to Martin.

The request had the backing of several influential entities, including the city of Martin, Weakley Co. Farm Bureau, Tennessee Farm Bureau and others, all of whom sent telegrams to the War Assets Administration.

The Spinks Co. saw building #136 as a prime location for a woodworking plant.[615] The U-T Junior College wanted the building for a stock judging pavilion and requested a "certificate of need" which would entitle it to the surplus property.[616]

As negotiations narrowed to the Spinks Co. bid, local newspapers got wind that the wrangling over building #136 had the potential for gumming up the entire land sale.

---

[614] Ibid.

[615] "Camp Tyson Ghost Town Days Are Over," *The Paris Post-Intelligencer*, June 5, 1947, front page, second section.

[616] Louis Newman, "The Camp Tyson Story," *The Parisian*, May 23, 1947.

"Camp Tyson could be a busy place today. Machinery could be moving in. Trucks could be threading through those deserted streets. The place, conceivably, could be humming," *The Parisian* pointed out on May 23, 1947. Noting that the certificate of need had been granted because of the U-T Junior College's veterans' program, *The Parisian* called on the powers that be to get the problem resolved post-haste. [617]

BID AWARDED TO SPINKS

As negotiations continued, the War Assets Administration (WAA) — probably to no one's surprise — rejected the Young Business Men's Club's proposal. The WAA noted that the club's bid did not cover all of the area up for sale "and, if accepted, would leave the balance of the installation in the custody of WAA." [618]

With that rejection, the Young Men's club endorsed the Spinks proposal, sending a telegram to the WAA at 5 p.m. April 26, and following it up with a letter from Brown Morris. In it, the club stated it believed "the bid on Camp Tyson, submitted by Spinks Clay Co., as fulfilling the aims of the club regarding prospective industrial development of community." [619]

At this point, negotiations for the property began to heat up. Paris Mayor Frank Blake (who had operated the city's bus line to Camp Tyson during the war) and Ed McClure of the Henry Co. Farm Bureau traveled to Washington, D.C., to push for the WAA to award the property to Spinks.

R.B. Carothers of Spinks dispatched his attorney Robert G. Jeter of Dresden, Tenn., to Washington, and told him, "Don't come back from Washington without a signed deed." [620]

"Heavy political pressure" was being brought to bear for the National Guard's acquisition of building #136, including State Adjutant General Hilton Butler of Nashville. [621]

But the Spinks contingent had friends in high places, too, including legislators. In its proposal for the property, Spinks also noted that R.B.

---

[617] Ibid.

[618] "Exhibit E, Comparative Analysis of Proposals," property of Rebecca Goins.

[619] Telegram, dated April 26, 1947, and follow-up letter, dated April 26, 1947, and received by WAA on April 29, property of Rebecca Goins.

[620] Carl Ross Veazey, telephone interview with author, August 11, 2009.

[621] *The Parisian*, May 23, 1947.

Carothers, Sr. was a Lieutenant who had served as officer in charge of construction of the submarine building yard at New London, Conn., and commanding officer of U.S. LST 692, and that his sons, R.B. Jr. and Harry Carothers each served in the Marine Corps.

Representative Tom Murray sent a letter to the WAA on May 7, 1947, endorsing the Spinks proposal and including telegrams in support from: C.H. Parks, President of the Farmers Bank and Trust Co., C.E. Hastings, President of the First Trust and Savings Bank; Brown Morris, A.C. Jackson, a leading Henry Co. farmer, Ed McClure, Mayor Frank Blake, Lee Dyer of American Legion Post No. 8, Jack Veazey, commander of American Legion Post No. 8, Richard Schofner, local merchant and Fleetwood Lowe, chairman of the Henry Co. Highway Commission, all of Paris.

LAYING IT ALL OUT

To facilitate the bottleneck, the Spinks Co. offered a proposal to the WAA. At a meeting on May 7, 1947, in Washington that involved U-T Junior College representatives, Carothers, Company Attorney Jeter, and Congressman Murray, the Spinks Co. offered to furnish a truck to haul another building to the college.

Laying out a map on the desk, Carothers explained how his company would utilize the Camp Tyson site. In an inner-office memorandum, J.E. Pilson, property disposal officer, wrote: "Mr. Carothers estimates that he will be in a position to employ immediately from 200 to 300 persons and as the project grows this employment would be increased; also, further, other industries would be attracted and we are assured that several other companies are interested in locating at this site, such as an auto parts and lime storage business." [622]

In the end, U-T Junior College in Martin did acquire one of the buildings from Camp Tyson, which ended up being used as the ROTC building on campus for many years. It has since been demolished.

Even with the negotiations now narrowed down to the Spinks Co., bargaining by the federal government continued. The bid by Spinks of $171,000 was rejected by the government. A counter-proposal by Spinks to purchase the property for $186,000 was offered and also was rejected.

---

[622] "Exhibit B," inner-office memorandum of War Assets Administration, May 7, 1947, property of Dickie Carothers.

On May 20, 1947, the War Assets Administration then authorized a counter-proposal of its own, offering Spinks the property for $201,000.

Although it was $30,000 more than it intended to pay, the Spinks Co. agreed. Attorney Jeter was able to return to Paris with a signed deed.

Ball Clay Deposits Draw Interest

Dickie Carothers of Paris, whose grandfather was R.B. Carothers, said his family became very interested in the property because of what it could offer for their company. Henry County is rich in ball clay, which is ideally suited for the ceramic industry.

The Spinks Co. was named for Harry C. Spinks, who was the father of Josie Carothers, the wife of R.B. Carothers. Spinks was in the forefront of the Tennessee ball clay industry and had begun mining deposits in Puryear (in Henry Co.), Gleason (in neighboring Weakley Co.) and other areas in western Henry Co. at the turn of the century. "Around World War I, almost all of the clay came from England for the ceramic clay industry," Carothers said, "but my great-grandfather decided it would be profitable to mine the clay deposits we had right here." [623]

In those days, mine excavation was a primitive endeavor and the company utilized wheelbarrows, horses and mules.

Spinks also was interested in cattle and brought the first herd of pure-bred Hereford cattle to Henry Co. in 1937. By the World War II era, Spinks had died and the company had passed to the Carothers brothers.

Mining techniques improved with the passing of the years, and Dickie Carothers, grandson of R.B. and Josie, recalled: "We had been mining ball clay, or ceramic clay, which has applications for toilets, sinks, sanitary sewers, plus other uses, for many years, but after the war, what the company really needed was an industrial site with a railroad spur and that property provided that." [624]

---

[623] Dickie Carothers, personal interview with author, August 3, 2009.
[624] Ibid.

Another attraction for Spinks was the 90' barrage balloon hangar "which we could use, also. The building worked very well as a reducing site and it gave us storage for the clay," he said. [625]

The Camp Tyson property already had a system of paved roads and a sewer and water system, he said.

That did not mean that the property did not need a lot of work. With the property standing idle for a year "the water pipes had deteriorated. The water and sewer systems were there, but not operating. There was so much iron in the pipes since it had been idle for so long," Carothers said. [626]

CONVERTING THE PROPERTY

The Spinks Co. set about converting the Camp Tyson property. Offices and motor pool buildings were converted to "offices, maintenance shops, labs," Carothers said. [627]

And the Carothers family had an ingenuous way in which to utilize the barracks buildings. Most were either destroyed or sold, but many were salvaged and converted for use as apartments in which company employees could live for free.

"After the war, there was a housing shortage," Carothers said. "So the company supplied its employees with housing. The Spinks Co. never charged rent and all utilities were paid." [628] That arrangement lasted through the 1950s, he said.

General Maynard's living quarters was converted into a home for R.B. and Josie Carothers and the other Carothers' family members also lived in officers' quarters on the grounds. With the Spinks Co. also engaged in the cattle business, many photographs depict cattle roaming free on the front yard of the Carothers' home.

A RUSH FOR JOBS

Once the Spinks company began its move onto the property, another rush of job-seeking associated with the camp grounds occurred. "Spinks Clay Company officials reported this week that during the first few days after moving into Camp Tyson, they had literally 'hun-

---

[625] Ibid.
[626] Ibid.
[627] Ibid.
[628] Ibid.

dreds' of applications for work," according to *The Parisian* newspaper. "Apparently recalling the lush days of the camp's construction, people flocked in to apply for jobs as carpenters, plumbers, electricians, or most anything you might mention." [629]

Accompanying a front page article on June 6, 1947, *The Parisian* published photographs of the ownership papers changing hands, along with members of the Carothers family, city officials, Attorney Jeter and officials from the War Assets Administration. Just a few hours after the camp was officially transferred to the Spinks Co., employees began moving into the former barracks on the grounds. First to move in were Mr. and Mrs. Cleatus Tyler of Puryear, with their children, Cleta Fay, 3, and Donna, nine months old, whose photograph also is included in *The Parisian* photo spread. [630]

The former Camp Tyson provided the Spinks Co. with an ideal headquarters. The clay drying and grinding plant was erected in the former balloon hangar, which was located adjacent to the offices and shops. "The plant was Spinks-designed and still reigns unique in the industry because of its controls, which permit maximum drying and grinding of clay without destroying plasticity. At the headquarters, the company also installed a laboratory for mining and plant control and for customer service." [631]

A cemetery is still on the site, one which the Army set aside for burials. A brick wall surrounds the cemetery, but it now is overgrown with weeds and saplings. Carothers said no graves were found in the cemetery when his family took over the property. [632]

The Carothers family dissolved its partnership in 1949, and many employees joined in the ownership as stockholders. The Carothers family members remained on the board of directors, as did other officers of the new corporation, including Carl Ross Veazey, who worked for the company for many years as assistant treasurer, then treasurer. He also was one of the many employees who lived on the site.

---

[629] "Sign Of The Times," *The Parisian*, June 6, 1947.

[630] Ibid.

[631] Untitled history of Spinks Clay Co., written by Don Bowden, n/d, property of Dickie Carothers.

[632] Carothers interview.

The Spinks Co. was known for loyalty of its employees. It is said that once you started working there, you stayed there, often for decades. And a big part of the appeal was the housing arrangements for its employees.

"It was a nice opportunity," Veazey said. "No rent or anything. And, it was real handy if Mr. Carothers needed you on Saturday or Sunday," he laughed.

Veazey and his wife lived in a former barracks for years. "Actually, our second son was born the day after we moved in there."[633]

Betty Hart's family also lived in former barracks while her father, J.L. 'Pod' Rawls, worked there tending cattle.

"We lived in a big barracks that was divided into apartments. Three different families lived in the barracks where we lived and even then, our apartment was a pretty good-sized. We had a big living room, bedroom, kitchen, bathroom, another bedroom, a long hall, and three more bedrooms, then a washroom. Pretty big."[634]

The roads were still in good shape then, she said, "and it was a lot of fun living out there. We played by the old water tank and we picked buttercups by the hundreds. There were the most beautiful japonica bushes out there." The Carothers and the employees "used to have barbecues out there every summer. Everybody really got along well."[635]

The school bus from Henry would pick up children inside the former camp, she said. "There were so many kids living in those old barracks, they had to. We lived there while I was in 5th, 6th, 7th, 8th grade. Actually, still through high school." Cattle roamed free "and you had to drive slow and stop for the cattle all the time," Hart said.[636]

The former fire stations also were converted to apartments and it was in one of those buildings that Jimmy Martin's family lived after living in a former barracks building for 20 years.

"The barracks we moved into had a main part, a long hallway and rooms on each side. Daddy knocked a wall out to make it bigger for our living quarters," Martin said.[637]

[633] Carl Ross Veazey, personal interview with author, August 11, 2009.
[634] Betty Hart, telephone interview with author, Feb. 15, 2010.
[635] Ibid.
[636] Ibid.
[637] Jimmy Martin, telephone interview with author, June 9, 2009.

Martin remembers playing there as a boy. "The big culvert things were still there and behind the old feed mill were the old ammunition bunkers and we used to play in them. There was an old rifle range and three bunkers were there, a Quonset hut. We used to play there, too." [638]

"Everybody used to live out there. Mr. Veazey and his family. Charlie McClure, he was the boss over the camp, the superintendent. He always had a dog, Monk, who rode in the car with him everywhere. If you wanted to ride in the car, too, you rode in back and Monk rode in front. Monk was buried by the old flag post." Fire hydrants were still on the grounds, a well house, and the huge water tank, he said. [639]

As an adult, Martin worked for the Spinks Co. until he retired. He recalls the bittersweet day when the barracks that his family had lived in was torn down. "A lot of those old barracks houses were really starting to get in bad shape and my boss said, 'We're going to tear down your old house. You can be the one who tears it down if you want to. So, I did. It was a pretty sad thing to do." [640]

"IT WAS LIKE FAMILY"

Ralph Anderson of Paris is another former employee who worked at Spinks for years: 48 years, in fact.

"It was actually like a little city out there," Anderson said. Anderson's family farm abutted Camp Tyson when it was still in operation "and as a young fellow, I'd go play on the camp. I remember playing on those whirling things by the sewer system." His sister married one of the Military Policemen who worked at the camp. [641]

As an adult, he began working at Spinks and recalls, "There were a lot of third, fourth generations working out there. It was a good company to work for. It was more like family."

CATTLE AND COWBOYS

Spinks was a unique place to work for another reason: where else in Henry County could you be a cowboy?

---

[638] Ibid.

[639] Ibid.

[640] Ibid.

[641] Ralph Anderson, personal interview with author, August 6, 2009.

One of the local men who was lucky enough to be in that position was R.L. "Shorty" Hutcherson, who started working on the farm out there in 1949. "They were going to lay off a bunch over Christmas, but Mr. Dick came by the garage and asked someone, 'What do you have this boy doing?' They said nothing. So, he told them to send me to the horse barn." [642]

And in the horse barn he stayed, for 45 years until he retired in 1994. Hutcherson had been raised on a farm, but had never ridden a horse before. "All Daddy had was mules. I'd never ridden in a saddle until I worked for Spinks," he said. [643]

The cowboys who worked at Spinks tended to the horses, rode the horses every day, shoed them, branded them, and showed them at the regular horse sales that were held there. Occasionally, they also rode the wild horses to tame them, which was a particularly dangerous part of the job. Hutcherson's main horse was Tennessee Stud. "I was the one who broke him, so he became like my horse," Hutcherson said. [644]

Veazey said one of his favorite things to do was hanging around the horse barn during slow moments. "I really enjoyed being around the stables. Mr. Dick would let me come down, especially when he needed help castrating the horses. He'd be looking for any available man to help." [645]

Richard Carothers Sr. "really enjoyed the horses. Cow Lady was his favorite mare. He just loved being around the horses I think more than anything," Veazey said. [646]

Veazey enjoyed helping brand the horses, too. "I enjoyed it, but some did not. I was the treasurer out there and there's actually a lot of bookkeeping involved in branding. Mr. Dick was so meticulous he branded everything. There were times I thought he was going to brand us." [647]

After Carothers died, the company dispersed most of the show horses and the sales were discontinued.

But several horses still remained there. "The horses were employees, like everyone else. They were used to check the fences around the

---

[642] R.L. "Shorty" Hutcherson, personal interview with author, April 8, 2010.
[643] Ibid.
[644] Ibid.
[645] Carl Ross Veazey personal interview with author, May 7, 2010.
[646] Ibid.
[647] Ibid.

property, corral the cattle," Veazey said. "Working there was a very unique experience. I really enjoyed it." [648]

Cattle and horses are not the only animals that have a history on the property, however.

Even before Spinks purchased the grounds, a unique use was found for it when the National Foxhunters Association selected the property for its 52nd annual Bench Show and Field Trials in November of 1945.

According to Bryant Williams' *Post-Mortems Vol. 2* (based on his columns in The Paris Post-Intelligencer), local excitement ran high, as every hotel room within miles had been booked solid for months. Fox hound owners and fanciers "came from all over the country to watch the hounds 'cast' early each morning in pursuit of the wily red fox," which roamed the area freely in those days. [649]

FALLING IN CISTERNS

Hilton Wygul of nearby Henry was in the fertilizer business and one of his biggest contracts was handling the Spinks property. His son, Jeff, used to work for the family business and, as a history buff, one of his biggest thrills was accompanying his father there.

"We did virtually all the lime and fertilizer work for Spinks and there was a lot to do," Wygul said. "I think I've been over every square inch of the pasture land out there at one time or another. I remember there was an old ordnance bunker right by the entrance at Gate 3 Rd. and all you could see was the door. The rest was underground. We also had to be careful when we were driving those lime trucks not to fall in the cisterns that were still out there. We did have a couple accidents doing that." [650]

Jimmy Martin recalls the underground bunkers, too, and said rumors were swirling around that the military had left ordnance or other material in them. "The bunker was all welded up and people were afraid there was something buried down in there, so we took a blow-torch to it and went down in there, but nothing was there." [651]

---

[648] Ibid.
[649] Bryant Williams, *Post-Mortems, Vol. 2*, page 44.
[650] Jeff Wygul, personal interview with author, November 12, 2010.
[651] Martin interview.

As the Spinks Co. continues operating under the name Lhoist North America Spinks Clay Co., having been acquired by the Lhoist Co. of Belgium, it still is in the forefront of the ball clay business.

And, cattle and horses still are on the grounds today. Horses continue to be cared for in a stable and corral area near the front entrance and cattle continue roaming free over the property.

The miles of roadways at the site are not in good shape anymore, with concrete breaking and weeds coming through the cracks. The large water tower that could be seen from Hwy. 79 (which for several years was called "The Camp Tyson Highway") was destroyed some 20 years ago. Dickie Carothers said it had deteriorated.

"It was a pretty big operation to bring that down," he said. "A salvage company we sold it to cut the legs and a cable was attached and it was brought to the ground. It was so full of rust, there was just a red cloud that went up in the air. It was pretty spectacular, actually." [652]

Gen. Maynard's old quarters are still on the site and the former Camp Tyson motor pool and office buildings are still being used for offices and laboratories. The 90' balloon hangar is still there and still used for drying and storing clay.

There are two entrances now: one which today is the main entrance and the other at Gate 3 Rd. However, because of vandalism on outer reaches of the site, the company now has begun shutting the entrances on the weekends.

What used to be the site of the guard house at the former main entrance at Palestine Rd. can still be seen, although with weeds have grown up considerably.

Present employees are well aware of the legacy of which they are guardians today. Chris Corley of Paris loves history and he, along with Manager Dan Collins have insured that boxes and boxes of old photos and documents are still stored at the site. Photographs of barrage balloons adorn the walls in the offices and lunchroom.

"We all really think it's important that as much of the past that was lived here is still preserved," Corley said. "I really love coming out here to work because I can really feel all the history that happened here, every time I come here." [653]

---

[652] Carothers interview.

[653] Chris Corley, personal interview with author, May 15, 2009.

Chapter 27

Soldiers Tell Their Stories

JAMES WILSON: IN A LEAGUE OF HIS OWN

THE late James Wilson of Paris was in a league of his own. Of the dozens of Camp Tyson soldiers who married local girls and settled in Henry County after the war, Wilson was the last surviving soldier.

When told he was the only Camp Tyson veteran left in the county, he sadly shook his head. "Oh, me," he said.

Wilson has plenty of memories and photographs of his Camp Tyson days to keep him company, as well as the woman who drew him to remain and make a life in Henry County, his wife, Anne.

Impressed with his rugged good looks, the Army brass compelled Wilson to pose for many publicity stills while he was stationed at Camp Tyson. In some photos, he is posing with barrage balloons, in others in front of barracks or the medical wing.

As with many Camp Tyson soldiers, Wilson found the camp was not fully constructed when he arrived. Wilson took a winding route to the camp, beginning in Fort Jackson, S.C., then on to Fort McClellan in Alabama for basic training, and finally to Camp Polk (now Fort Polk), La., where he was assigned to the medical detachment.

"They asked us if anybody had any skills," he recalled. "Well, I had taken typing in high school, so I was assigned to be a clerk for the medical supply. I typed index cards for the medical warehouse."

He was at Camp Polk when the attack on Pearl Harbor occurred and that accelerated activity for him and his fellow soldiers. "I was in Shreveport on a weekend pass on that Sunday. We were on the edge of town, thumbing back to camp, and people stopped along the road and told us about it," Wilson said. "Of course, I was upset and concerned, but not afraid."

When he reached the camp, he said, "They were already beginning to ship the armored division out."

It was then that he learned he was one of 20 other men in his unit who would be sent to Camp Tyson. They had never heard of Camp

Tyson and likewise had no idea the camp would be a barrage balloon training center.

"When we got to the camp, there was tremendous work going on; the civilians were building the camp from the ground," he said. "I can vividly remember the red gravel streets."

Wilson was again assigned to the medical detachment, but the hospital had not been built yet. "There were actually only four barracks built when I got there, plus the headquarters for the quartermaster. And we had a first aid station where you could go on sick call, but it took about three months for them to build the hospital."

The equipment for the hospital began arriving "and every day we helped install it," Wilson said. "We had 19 wards when it was all done, plus a headquarters, nurses' quarters."

In the early days, "there really wasn't much to do," he said. "But all the time, there were more men coming in to the medical detachment. Officers and doctors began arriving."

Eventually there were 325 enlisted men assigned to the medical detachment.

"We had three men on duty 12 hours a day in each ward," Wilson recalled. "Col. Blackwell was the commander of the hospital and we had routine duty every day." The hospital included a medical ward, lab, pharmacy, surgery, and a sick call ward.

"We reported to the sick call ward every morning. There always a group from all over the camp that had reported in sick," Wilson said.

The camp was operated like any other Army camp, with regimented daily routines, he said.

"At 6:30 a.m., we had reveille, chow at 7 a.m., then we'd go to work. We did reveille out on the street for each barracks and someone would call the roll. The officers of the day were assigned and if anyone missed reveille, they'd report to the orderly room."

Work ended at 5 p.m., when the soldiers would stand at retreat, "the guns would go off and we'd stand at attention until they lowered the flag," he said.

Work for Wilson was in the medical supply section, working in the various warehouses. He was soon transferred to the adjutant's office as his assistant. "I put my typing skills to use again," he said, "and sometimes I would drive the officers."

Wilson was promoted to Staff Supply Sgt., and then to First Sgt.

There were five barracks for the medical detachment, he said, including offices, medical supply, morgue and Red Cross.

"Actually, the medical detachment was like its own camp; we were kind of apart from the other part of the camp. We had a dental clinic, eye clinic, mental ward..."

Wilson was performing the detachment's weekly laundry run when he spotted his future wife working in the office of the laundry. He asked her out and they went on their first date to see "Gone With The Wind" at the Gem Theater in Paris. They began dating steadily and married in 1942.

Wilson said there was plenty to occupy soldiers at Camp Tyson, from the USO in Paris, to dances at the camp, to the sports teams that were organized there. He played baseball and basketball with the camp teams and said, "I enjoyed that; we got to go to different places and play, so I got to see a lot of the area that way."

Wilson stayed at the camp until the summer of 1944, but his service to his country was not over. From there, he was sent to California to be shipped overseas. "First we went to New Guinea, then to Luzon in the Philippines, and then on to Okinawa, Japan. We were there when the atom bomb was dropped on Japan," he said.

When the war was finally over, he was dispatched to Seoul, Korea, to serve with the army of occupation. He was discharged from the Army in November 1945.

Upon his return to Paris, he was encouraged by his father-in-law, Judge Jim Thompson, to accept a position at the Kentucky-Tennessee Clay Co., which became his lifelong career. He became manager of the Tennessee division of the company and worked there for 38 years. He was active in civic duties, serving as magistrate in the old County Court and a county commissioner.

WILLIAM DABNEY: LEGION OF HONOR WINNER

William Dabney said he was "too afraid to be afraid" at the D-Day invasion at Omaha Beach on June 6, 1944. "I was too excited to be scared. I was just trying to save my life. I was too alert to be scared. I was too busy trying to take cover."

Dabney was recalling the hellish invasion and the role that he and other members of the 320th Anti-Aircraft Barrage Balloon Battalion played in it. The 320th was the only all-black regiment to be on the beaches of Normandy on D-Day.

The 320th was also one of three all-black units that were trained at Camp Tyson. The other units were the 318th and the 319th.

Dabney was stationed at Camp Tyson for 1 ½ years, he said, learning how to wield and properly use the huge barrage balloons. From Camp Tyson, the train took the men of the 320th to New York and then boarded a ship for England, where they received six months of combat training. "We already were trained on the barrage balloons, but we needed training for combat, for being under fire," Dabney said.

As June drew near, the men then boarded a ship to cross the English Channel "and we knew we were going to take part in the invasion. General Eisenhower came on the loud speaker on the ship and said, 'Some of you will not be returning.' That got you upset a bit, hearing that."

Once their ship landed at Omaha Beach, members of Dabney's battalion were right behind members of the 29th Texas Rangers. "We weren't the first ones off the ship, but we were right after the Texas Rangers," Dabney said.

What made Dabney and the other members of the 320th unique from the Texas Rangers and all the other battalions at D-Day, was that they were attached to the heavy barrage balloons as they came off the ship. "I had my balloon attached to my belt. I remember flying in the air from the strength of the balloon when we came off the landing board."

Dabney said he weighed 170 lbs. at that time "and I had a 25-pound pack on my back" and even with a smaller balloon like he used for the D-Day invasion, the heavy balloons thrust the men in the air. The balloons were designed "so they wouldn't lift us too high off the ground."

Once on the beach, he said, "just about the time I got off the landing board, the balloon was shot out from under me. I don't know if it was a plane that hit it or if the Germans shot it down, but it was shot down and I immediately disconnected the cable. I threw myself on the sand with my crew."

Dabney and his crew spent two days on the beach, until General Patton and his tank division arrived. Four members of the 320th were killed in action on D-Day.

In the photographs of the D-Day invasion, one of the more memorable are the barrage balloons that were seen floating in the air, amid the smoke and the battleships along the shoreline.

And yet, the men who operated the barrage balloons were all but forgotten for the past 65 years, until the men who served in the battalion began being recognized for their valor. Dabney is the one who has received the highest honor, the Legion of Honor, which was awarded by the French government on June 5, 2009.

"I felt pretty proud" about the honor, Dabney said. "I wondered if it would ever come in my lifetime."

At the ceremony, Dabney said he sat directly behind President Barack Obama, French President Sarkosy, British Prime Minister Gordon Brown and Prince Charles. "When it was over, I went up to shake President Obama's hand and he saw me and said, 'I know you, I read about you in *The New York Times*.' Then his wife, Michelle, saw me and she said the same thing. She said she was so proud of me and she kissed me."

When Dabney returned home a couple nights later, his wife told him he would probably never wash his cheek again, he said.

For Dabney and the others, the recognition they are now receiving is bittersweet. He was interviewed for the History Channel documentary "A Distant Shore: African-Americans At D-Day," which is now on DVD.

Dabney also was the focus of several newspaper articles, including one in *The New York Times* written by Brian Knowlton, and others published by *The Roanoke Times* and The Associated Press, which were distributed nationally. [654]

"I would have thought they would have recognized us before now, after all, it's been 65 years," Dabney said in 2009. He noted that it was especially galling to see "Saving Private Ryan" and not see any black soldiers or barrage balloons in its depiction of the D-Day invasion.

Looking back on his war years, Dabney said it wasn't bravery that led him to enlist. He was 17 and living in Roanoke when he enlisted, he said, but he did so because his older buddies had been drafted "and I didn't have any buddies to hang out with." Because he was under-age, his grandmother had to sign the papers allowing him to enlist.

Even after the trauma of D-Day, Dabney's war activities were not over. "I came home, to Camp Shanks, N.Y., and got a 30-day pass. Then we went to a camp in Georgia and I thought I was going to be discharged, but instead I was sent to the Philippines. We were in international waters when we heard the war was over." Dabney was dis-

---

[654] Joe Kennedy, "Member of U.S. Army's 320th Anti-Aircraft Balloon Battalion Returns To Omaha Beach," *The Roanoke Times*, June 16, 2004; Sue Lindsey, The Associated Press, reprinted in *The Washington Times*, June 7, 2009; Brian Knowlton, "Blacks' D-Day roles lauded on anniversary," The New York Times, June 6, 2009; Shannon McFarlin, "Former Camp Tyson Soldier Receives Legion of Honor For D-Day Valor," *The Henry Countian*, June 9, 2009.

charged on Nov. 15, 1945, after having been in the service for 2 years and 11 months.

In civilian life, Dabney earned a degree in electrical engineering, but could not find a job in that field. He worked for flooring companies, learning the trade, and opened his own business, Dabney Floor and Tile Service, which he operated for 37 years. [655]

EDDIE CLERICUZIO: TRAINED A LOT, MARCHED A LOT

When he arrived at Camp Tyson, it was a shell of what it would become. As a member of the 1284th Combat Engineers, he was in the first platoon to arrive at the camp while it was still a muddy, unfinished mess.

As related in Chapter One, Clericuzio and his fellow soldiers were required to clean the sawdust off the floors and set up their own cots before they could go to bed their first night on base.

"The barracks still smelled like new. It had just been finished and sawdust was still on the floor," Clericuzio said. Clericuzio was 26 years old when he was drafted, older than most of his comrades in arms, and was already a sergeant by the time he arrived at Camp Tyson from Fort Dix, New Jersey.

From a few buildings, the camp began to steadily build and grow into its massive size during the three years that Clericuzio was stationed there. "We trained a lot; marched a lot," he said, noting he was as fascinated with the barrage balloons as most everyone else. "But everything was a secret out there. They didn't tell you anything. They'd build the balloons in the big hangar out there, inflate them and use a jeep and a winch to get them out of there. Then they'd ship them out to wherever they were needed."

Clericuzio grew up in an Italian family in Bloomfield, New Jersey, and his language skills came in handy when Italian prisoners were brought to the POW camp at Camp Tyson. (See Chapter 23). He said he "got along pretty good" with the prisoners, who were happy that someone there could speak their language.

While at Camp Tyson, Clericuzio learned a skill that would keep him busy for the rest of his life. "They had a big, nice theater on base and I liked to run the movies on the projector. I really like working

---

[655] William Dabney, telephone interview with author, June 9, 2009; Shannon McFarlin, "Former Camp Tyson Soldier Receives Legion Of Honor For D-Day Valor," *Henry Countian*, June 9, 2009.

with the projector and got to where I could take it apart and put it back together." After the war, he returned to the neighboring town of McKenzie, where his local bride was living, purchased the Park Theater there and continued operating it for 37 years.

"I met Carolyn Thompson while I was at Camp Tyson. She was born and raised in McKenzie, so that's where we lived after the war," he said. "When I heard I was going to be shipped overseas, she said she wanted to get married before I was shipped out, so we married in July 1945."

Clericuzio first was transferred with the combat engineers to a base in Texas, and then on to the European theater, where he "saw a lot of combat" in Belgium, France and Germany. "I was just glad for the war to finally be over and be able to be home." [656]

HOWARD KOENEN: HELPED LAUNCH THE FIRST BALLOON

Howard Koenen of Connecticut was drafted way before Pearl Harbor, in August of 1941. He was first sent to the reception center in a camp in Massachusetts "and I stayed there so long I thought I'd never leave. I could go home on weekends, so it wasn't really like being in the service at all."

He finally was dispatched to Camp Wallace near Houston, Texas, where he was called upon to put his experience as a mechanic to use. His military days were spent rather idly, but "then came Pearl Harbor," Koenen said.

"We were hitchhiking that day," Koenen said, "and someone picked a group of us and took us into Houston. We wanted to go to the Midget Auto Races. We were hardly seated when the public address system announced that all military personnel must go to their stations."

When he and his buddies returned to the camp "things were different on camp. The guards had live ammunition. They'd never even used guns before. I remember that first night; they started sending things out of the camp."

Anti-aircraft guns were stationed around the oil refineries for "civilian morale," Koenen said. "They were deathly afraid the Japanese were going to land around the Gulf of Mexico."

Koenen spent eight weeks at Camp Wallace and was sent to Camp Davis in North Carolina, where the barrage balloon school was tem-

---

[656] Eddie Clericuzio, personal interview with author, May 11, 2009.

porarily located. "They sent out orders to move to Tennessee," he said. "My buddy had a 1933 Willys and he didn't want to leave his car. He said he could probably finagle the officers into letting him drive it, so that's what we did. I drove the car and we stayed with the convoy driving from there to Paris."

Koenen was assigned to the 302nd Barrage Balloon Battalion, Battery B. When he arrived in camp in early 1942, much of the camp, except for his barracks, was still being built.

"It was our battalion that put the first balloon airborne at the camp," on Feb. 13, 1942. It was quite an accomplishment," Koenen said. His job was to maintain the trucks, "but I observed them putting the first balloon up in the air." He said there was a "big celebration" that day.

Koenen is one of the several men who found their brides while at the camp. He and some buddies had driven into Paris and saw some girls on the side of the road with car trouble. Koenen helped fix their car and ended up with a girlfriend, a girl named Neva from Murray, Kentucky, which is 22 miles from Paris.

He and his wife married November of 1942 and they took an apartment in Paris while he continued serving at Camp Tyson.

With the use of barrage balloons winding down, Koenen said, "they started sending the privates out, then the corporals, then the sergeants and our outfit was still there. The equipment was taken out and the barracks were empty. We stayed in an empty barracks and finally one day I asked what they were going to do with us."

An officer took him to the motor pool and pointed to the bulletin board. "The orders said that the 302nd would proceed to Camp Breckinridge with all its organization and equipment. That meant everything you have that's been issued by the government. And they really meant it. It included cots, everything."

The soldiers "cleaned everything out, loaded it on the trucks. We started out and proceeded through Murray. I remember seeing my wife standing on her parent's porch on 12th Street and waving as we went by."

The 302nd arrived at Camp Breckinridge, but had to pull the convoy off to the side of the road. "They don't have word that we're supposed to be here yet, they told us. After a day or two, we went on in," he said.

His wife moved to Camp Breckinridge to be with him and they rented a room in a doctor's house. "I remember she helped him with a tonsillectomy one time — she had to hold the patient's head for him," he laughed. Koenen was promoted to First Sergeant while at Camp Breckinridge.

After the war, the couple first moved to Connecticut, then back to Murray, where he opened a mechanic shop and she worked as a beautician. They had two daughters, Cindy and Kathy. They were married for 60 years, until his wife died in 2002.

Koenen was active in civic affairs, serving 23 years on city council and other duties. "On my 90th birthday," he said, "I quit everything." [657]

WILSON CALDWELL MONK: AT THE D-DAY INVASION WITH THE 320TH

Wilson Caldwell Monk of Atlantic City, N.J., was drafted in 1941, and initially stationed in Fort Dix, New Jersey, then on to Fort Eustus, Virginia, for basic training. From there, he was dispatched to Camp Tyson in early 1942, "where we formed the 320th Anti-Aircraft Barrage Balloon Battalion."

Like so many others, Monk arrived by train, finding a still- unfinished camp.

His cadre was put to work learning the ins and outs of balloon operation. "I was promoted to Master Sargent and was the balloon inspector for the entire company. I inspected each balloon site to make sure it was rigged, that the grounds were maintained properly and that the balloons were inflated properly."

The balloons, he said, were filled with hydrogen and each battery worked with its own balloons.

"It was a novelty, I have to say," Monk said. "It was different, interesting."

Many of his unit's lessons were taught inside the huge balloon hangar on the camp. "We were using the VLA balloons, the Very Low Altitude balloons that we got from Britain."

Operating the balloons "took a combination of technical and muscle power," he said. "It wasn't really complicated. Every man had a job to do."

After training for months, the men in the 320th were deployed to Camp Shanks, New York, where they were taken to New York City and boarded ships to cross the ocean to Scotland. "While we were there, we were constantly training, every day. Then we were sent to

---

[657] Howard Koenen, personal interview with author, October 30, 2009.

southern England, where we were billeted in Abersachen, a little town. Then we went on to Cardiff, Wales, then Southampton, where we boarded a ship and crossed the English Channel."

The 320th was not told what to expect or what they would be doing, he said. "We were just told to get ready, get in line and follow me. I can faintly remember hearing Eisenhower telling us about the invasion."

The men were anxious, he said. "There's always a certain amount of fear of the unknown. We certainly weren't going to a tea party."

Monk said he doesn't have "a lot of memory of that day" of the D-Day invasion at the Normandy Beach. "I remember the beach and remember flying the balloons. A lot of the rest is a blank and I'm glad because it wasn't a pleasant thing to see."

After D-Day, the unit stayed in France for a "brief period," which also included a lot of training. "Then we got word we were being shipped back to the U.S. The word was we were going to be used in the Pacific War."

The unit "bundled up everything and went back to the States. We got a 25-30 day leave, then reassembled back at Camp Stewart, Ga." They were shipped to Fort Lewis, Washington, and then to Hawaii, he said. "We were supposed to go on to Okinawa for an invasion there, but we stayed in Hawaii for eight months. That was the best thing that ever happened to me. I remember being happy and getting to do a lot of swimming there." [658]

LOYAL WHITESIDE: A LOT OF PLEASANT MEMORIES

If there is anyone who has good memories of his days at Camp Tyson, it is Loyal Whiteside. And if there is anyone who has documentation of his days at Camp Tyson, it is Loyal Whiteside.

Whiteside, who grew up in Mars, Pennsylvania, had just graduated from high school in 1941, when he was called to service. He arrived in Camp Tyson shortly thereafter and was assigned to the 317th Battalion, Battery B.

The ways of the South were rather foreign to him, as were his duties learning how to operate the barrage balloons.

Whiteside kept a scrapbook while at Camp Tyson, which is chock full of photographs, notes, programs from events that were held at the

---

[658] Wilson Caldwell Monk, telephone interview with author, April 13, 2010.

camp, programs for worship services, matchbooks, bus ticket stubs — you name it, he saved it.

The scrapbook presents a fascinating picture of his time at Camp Tyson, where he first encountered Southern girls and Southern cooking. Photos in the book show Whiteside and his buddies posing in downtown Paris, on the steps of the Post Office, at the barracks (with a pet dog), and on maneuvers his unit took elsewhere in Tennessee and Alabama.

Whiteside listed the men in his unit, along with humorous comments about each. He also commented on the food he had eaten. Cornbread, eaten at Camp Tyson: "Didn't like it so good." Sweet potato pie, at Camp Tyson: "Have eaten better." Corn fritters, at Camp Tyson: "Not bad." K-rations, on maneuvers: "Horrible." C-rations, in Alabama: "Not bad."

In the scrapbook is a program for church services for the 302nd Anti-Aircraft Balloon Battalion, on March 12, 1944, a program for a follies presentation that the 317th presented at the Post Theater #1 on June 28, 1943, and commendations signed by General Maynard which document Pvt. Whiteside being given his PFC stripe for outstanding military appearance and bearing.

He also kept newspaper articles showing members of the 317th awarded trophies for basketball and an honor plaque for having "the best mess in Camp Tyson." Other articles saved include ones relating how far and wide various barrage balloons landed after getting loose at the camp.

After the war, Whiteside settled in Wildwood, Fla. [659]

MAJOR WILLIAM S. JONES: TAKING ONE STEP FORWARD

Major William S. Jones of Fredericktown, N.J., entered the Army at the Customs House in Philadelphia, Pa., two days after Pearl Harbor, on December 9, 1941. He was deployed to Fort Meade, Maryland, that same day. The nephew of Cordell Hull of Tennessee, who was head of the Department of State, was battery commander for his unit.

He was allowed to go home a couple days for Christmas, and upon returning to Fort Meade, "We were asked if anyone would to like to join the artillery to take one step forward. I did."

---

[659] Loyal Whiteside, telephone interview with author, September 16, 2010; numerous email communications.

Jones "was immediately transported to Galveston, Texas. We were there but a short time when they asked if anyone wanted to go to Tennessee and serve in a barrage balloon unit. I had enough of Galveston, so I took the ole one step forward and arrived at Camp Tyson. I believe this was sometime in February or March, 1942." He was placed in the 308th Coast Artillery Barrage Balloon Battalion.

Jones got a visit from his mother, a friend of the family and his girlfriend that Easter. "While they were there, I took my girlfriend and we drove to a schoolhouse high on a hill, where we looked on the lights of Paris." That would be Grove High School, situated on the highest point in Henry County.

One of his most vivid memories of Camp Tyson, he said, was this: "Lights were out at 10 p.m. Shortly after that, a freight train would pass by the camp, blowing its whistle, a high note, a low note, then another high note. That whistle would reverberate amongst the hills so that you would hear it, then 6-7 times. It is occasion as that that gives us older people some of the pleasant memories of our past life."

Jones has another memory of a less sentimental nature. He was on KP: "That's kitchen police, but don't let the name fool you. Police work had nothing to do with it. You peeled potatoes, scrubbed the floor, washed the dished and so forth. We had steel cots. If you were to be on KP the next day, you draped your towel over the iron bar at the foot of the bed."

He got up in the middle of the night and moved his towel six or seven cots down the line. "We had a mess sergeant who was of Greek extraction. Having already awakened the fellow whose bed I draped the towel, he and others were unhappy because the mess sergeant was grabbing various soldiers by their feet and shaking them, saying 'Jonas, Jonas, Jonas' instead of Jones. Of course, he woke nearly all of us and me laughing, but admitted I was Jones and of course, saying that someone else had moved my towel."

Jones also had his first taste of 'white lightning' at Camp Tyson (See Chapter Eleven).

After Camp Tyson, the regiment was taken by train to Olympia, Washington, where they were trucked to a little town, Chico, where they stayed in pup tents in a schoolyard, bathed and shaved in a trout stream next to a school house.

"I do not remember how many balloons we flew up at Washington State, but I do remember we lost nearly all of them during a large storm the Canadian Royal Police failed to warn us was on the way," he said. "The wind would make the balloons dive and when they turned to go up again, their multi-strand cable would form an 'S' and snap."

He put in for a transfer to the infantry for 10-11 months in a row and he finally came through "and I reported to the Seventh Infantry Division at Fort Ord, California, on March 23, 1943. "We landed in the Aleutians on May 11, 1943. But that's another story." [660]

---

[660] Major William Jones, letter to author dated June 25, 2009.

Chapter 28

"Hi There, Soldier!"

WITH an elevation of some 620 feet, Grove Hill in Paris is the highest spot in Henry County. From there, you can see the shining lights of Paris and the surrounding area when darkness falls.

It is made for romance.

For Sue Edwards Wagner and Loyal Whiteside, it was the spot that spurred their wartime romance. Wagner and Whiteside found each other again in their 90s, for the first time since 1944. Both had lengthy, happy marriages with others in the ensuing years, but their memories of their short romance lingered.

Their first face-to-face visit in 66 years sparked talk of marriage.

Like so many others, Whiteside and Wagner were caught up in the excitement of the Camp Tyson days, when chance encounters could lead to enduring friendships.

A soldier in the 317th Barrage Balloon Battalion, Whiteside was standing near the statue of a Confederate soldier on the court square in Paris when a car pulled up. Wagner said, "Hi there, soldier!" and asked Whiteside if he would like a ride.

"I keep kidding her that she picked me up," Whiteside said during their September visit at Wagner's home in Memphis.

Wagner said, "I did pick him up!"

The two drove around Paris, finding their way to Grove Hill, which houses the E.W. Grove School. "We did do more talking than necking," Whiteside said, "and we got to know each other up there. And I must say, we didn't have the only car up there, either."

The spot on Grove Hill, he said, "was just beautiful. The whole town was lit up and you could see it all from up there."

Whiteside had a childhood sweetheart in Mars, Pa., but was captivated by the vivacious girl who had grown up a few miles outside of Paris, in the small town of Whitlock. And Wagner was captivated with him, too. "He was very handsome," she said.

After the war, Whiteside's sweetheart, Marilyn, became his wife. "When I was nine years old, I said I was probably going to marry her," he said. They had a happy marriage with four children until Marilyn died a few years ago.

Wagner met another Camp Tyson soldier, Hugh Wagner from Appleton, Wisconsin, and romance bloomed with a fierce exchange of letters between the two. "We actually only had two dates," Wagner recalled. "Our last date was when he told me he would be shipped overseas."

Her romance with Wagner was not serious until they began writing letters to each other after he was deployed to Europe. A medic, Wagner was stationed in France and Belgium and stayed with the army of occupation after the war ended.

Ironically, both Whiteside and Wagner's paths crossed, although both were unaware of it, at the Buchenwald concentration camp where both were deployed when the camp was liberated by American forces after Germany surrendered.

"Even after all the fighting, all the mustard gas, everything," Whiteside said, "seeing what was in Buchenwald was the worst experience of the war. And with Hugh being a medic, I'm sure it was for him, too."

For Whiteside, the war became very serious after his barrage balloon training at Camp Tyson was over and he found himself deployed overseas. "First we were sent to Fort Rucker, Ala., and then Camp Shelby, Miss., and our outfit was changed to the 93rd Heavy Mortar Battalion."

The unit learned to shoot mustard gas shells, he said. "The Army thought the Germans would resort to gas and we were trained so that we could fight back," he said. Eventually, he was sent to the Pacific for the expected invasion of Japan, but the two atomic bombs were dropped and the war was finally over.

It was then that he was deployed to Buchenwald and found the horrors there. "The Germans had kept Jewish prisoners, gypsies, Jehovah's witnesses, Russian prisoners there. What we found there was just indescribable."

Arriving in Pennsylvania after the war, Whiteside married and eventually became a studio photographer and his wife was a music teacher.

With only few dates before he was deployed, Wagner wasn't even sure she would recognize her future husband when he arrived back in the states. "It's funny," she recalled. "I never wanted to get serious and I wasn't worried about getting serious with Hugh at all. I thought,

As If They Were Ours

well, he's from Wisconsin, I don't have to worry about him, but he's the one I ended up marrying."

Wagner moved with him to Appleton with no hesitation, she said. "I loved it dearly up there. I loved the winters. We had a wonderful life. We were poor but happy." Her husband was a salesman, she said. "He was a super-duper salesman. He sold me!"

Her husband died in 1985, and she found herself not as able to cope with the bad winters alone. "My son was transferred to a job in Memphis and I moved there to be closer to family," she said.

Whiteside and his wife, Marilyn, visited Paris in 1993, "and I made an attempt to look Sue up, but heard she had moved to Wisconsin."

Years later, after his spouse died, Whiteside was able to track Wagner down with the help of the internet. "I was always fond of Sue and she kept popping up in my mind. I want to make clear that both Sue and I had wonderful marriages to wonderful people. I had one of the most wonderful wives in Marilyn and she had one of the greatest husbands in Hugh."

But after talking over the phone, he said, "We found that the spark had never left." From his new home in Wildwood, Florida, Whiteside traveled to Wagner's home in Memphis for a week-long visit in September, 2010. "We haven't left the house all week," Whiteside said. "We have so much to catch up on."

The two spent the time perusing his extensive scrapbooks he kept during his Camp Tyson days, complete with photographs, newspaper articles and mementoes, "and talking, talking, talking," he said.

Their romance was rekindled and the two were even discussing marriage, but sadly Wagner died before the two could take that step.

A contemplative Whiteside said after Wagner's death, "We were happy to have found each other and happy the way things were." [661]

---

[661] Email communications and letters from Loyal Whiteside beginning in November 2009; telephone interview with both Whiteside and Sue Wagner on September 16, 2010; email communications with Whiteside in 2012.

Chapter 29

Keepers Of The Flame

FOR many Henry Countians, the name Eddie Moody was synonymous with Camp Tyson.

A former civilian employee at the camp, Moody's time there made such a lasting impression on him that he pledged to keep the memory of Camp Tyson alive.

For years, Moody had been collecting memorabilia on the camp, which he kept in the front den of his house. Framed photographs of barrage balloons and soldiers hung on the walls, while items from the camp, such as a 'manometer,' were on display. Moody was only too happy to show his visitors his collection and enjoyed explaining what each of the items were. A manometer, by the way, is a gauge used for the balloons, which utilized oil or water to determine the balloons' pressure.

After World War II, Moody built a successful career as a realtor and together with his long-time wife, Betty, raised a family. But his days at Camp Tyson remained a highlight for him.

When still nothing more than a youngster, Moody drove a gravel truck when the camp was being built, hauling four loads a day to help build the roads there. Moody married Betty Carr on March 29, 1942, and headed to Detroit, Michigan, seeking job opportunities.

He worked for a time at a defense plant there and injured his back, moving back to Paris. At home, he discovered that Camp Tyson was in full operation and he found another job there, as a plaster helper.

Moody received his draft notice and entered the U.S. Navy. He became a first seaman before re-injuring his back and again returned to Paris.

With his Navy service experience, he was able to acquire a position as a civilian guard at Camp Tyson, serving as an auxiliary military policeman. He explained that veterans could be sworn in to the Army in that capacity, but their jobs were limited in that their authority did not extend beyond the camp.

The days at Camp Tyson were bustling, he said. "There was always someone coming or going," as he helped man the guard posts at the camp's entrances.

His happiest times at the camp, however, were when he was asked to chauffeur General John Maynard to and from Paris. "I used to haul General Maynard," he said. "They used me to bring him to town and carry him through camp because I was local." General Maynard, Moody said, "was one of the finest men I've ever known."

Years after the war, Moody and several former soldiers at Camp Tyson decided the time was right to get together again and the first of several reunions were held. Moody was chairman of the reunion committee from 1973-1992.

Others involved in planning the reunions included Bill and Winnie Perkins, Cedric Knott, Harry Wheat, Bobbie Jean Freeland, Cal Orris, Jerry Gammon, Brig. General Roland Parkhill and Tom Lonardo.

Former Camp Tyson servicemen from all over the country were drawn to the get-togethers, which also included public tours of the former camp grounds. Knott served as tour guide for many of the tours, offering humorous anecdotes about camp life.

Probably the largest of the reunions was conducted on April 17, 1992, at the Paris Elks Lodge. On that occasion, photographers were on hand to take photos of the entire group and individuals. Wayne Webb of Mansfield was on hand to make copies of the photographs of the camp that attendees brought with them: a treasure trove of photographs that he still has today.

One of the speakers at the reunion was Susan Gordon of the Tennessee History Society, who informed the group of the Society's "Home Front" project, for which several of the attendees were interviewed. *The Commercial Appeal* in Memphis also covered the reunion and published an article on April 19, 1992. [662]

As fewer and fewer former soldiers returned to the reunions, the events became a memory — but a pleasant memory for Moody. "Sometimes I feel like I'm still at Camp Tyson," he said. [663]

[662] Laura Coleman, "Extinct Army Base Lives In Memories Of Love and War," *The Memphis Commercial Appeal*, April 19, 1992.

[663] Shannon McFarlin, *The Paris Post-Intelligencer*; Deborah Turner, "Eddie Moody Recalls Camp Tyson," *The McKenzie Banner*, November 7, 2001; James "Spider" Dumas, "Camp Tyson Days Recalled," *The Paris Post-Intelligencer*, April 16, 1992; Kelly Dodd, "Veteran Stresses Importance of Defense," *The Paris Post-Intelligencer*, April 3, 2002.

George Davison, one of the heroic men of the 320th Barrage Balloon Battalion, kept memorabilia from his time at Camp Tyson and in the 320th — newspaper articles, banners, a flag, ticket stubs. He also wrote down his memories of his experiences and put them in a box.

His son, Bill, found the scrapbooks after his father's death in 2002. Looking through the mementoes and reading his father's diary spurred Davison to act.

"I wanted to make sure these men were not forgotten," Davison said.

Davison, of Waynesburg, Pa., has devoted much of his time to drawing attention to the 320th in a variety of ways: by starting a website devoted to the men, writing politicians to encourage them to recognize the 320th, and working with film makers and news reporters.

He has worked closely with Alice Martine-Mills of Paris, France, who Davison credits with helping to facilitate the award of the Legion of Honor to William Dabney in 2009.

"Dad's diary is what got a lot of this going," Davison said.

Like Dabney, Wilson Caldwell Monk and others, Davison was irked that the movie, "Saving Private Ryan" did not depict the black soldiers who played such a crucial role in the D-Day invasion.

In a letter to Congressman Charles Rangel of New York, who had worked to recognize the Tuskegee Airmen, Davison asked him to work as hard to recognize the 320th. "Before my father died in 2002, I promised him I would tell the story of the 320th as best I could and I have been successful," Davison wrote. "However, the story of the 320th needs to be told more and more. The men...deserve recognition for their service and their sacrifice during World War II."

Davison has been compiling a list of members of the 320th from various sources, including those men who were killed or wounded in action at D-Day.

"We have made a lot of progress in getting the message out," Davison said. "But there is still much more work to be done." [664]

---

[664] Bill Davison, telephone interview with author, September 8, 2009; numerous email correspondence.

This scene featuring Williams Street in Camp Tyson, shows how the living quarters within the camp were configured. (Author's collection)

A large crowd was on hand at Camp Tyson for its Army Day in 1942, which was the first chance the public had to see the camp once it was completed. (Joel Summers)

Among the crowds at Camp Tyson's Army Day in 1942, were these couples from McKenzie, Tenn: From left, Margaret Adams, Russellene Hilliard Summers, R.B. Summers and Martha Adams. (Joel Summers)

The Camp Tyson Fire Department in 1944. From left, Hubert Hooper, Joe Moser, Eldrigh Cathey, Pape Tucker, Harry Hicks, Richard Cronk and Bill Argo. (Argo family)

The last day that Camp Tyson was open and these were the staff which closed the doors for the last time. From left, Lyndon Shelton, Jeanne Townsend, unknown soldier, Mary Jo Burke Wyatt and Soldier Grissom. Townsend could not remember the middle soldier's name or the first name of Grissom. She said the last day was very sad and they left a black wreath over the door when they left. (Jeanne Townsend)

As If They Were Ours

James Wilson of Paris is front and center with a huge barrage balloon in this publicity photo taken by the U.S. Army. Wilson was in the medical detachment at Camp Tyson and was selected for many Camp Tyson photos by the Army because of his rugged good looks. (James Wilson)

Camera crew from the U.S. Army at Camp Tyson, photo taken March 9, 1943. (State Library and Archives)

James Wilson of Paris at the entrance to the medical detachment offices on Camp Tyson, along with his fellow soldiers. (James Wilson)

Very Low Altitude (VLA) balloons protecting troop trucks during maneuvers along the countryside. The photo was used on postcards sold at the Camp Tyson PX. (Author's collection)

Old Camp Tyson postcard depicting barrage balloon with tethers, taken in 1943. (Author's collection)

A battalion of Camp Tyson soldiers running maneuvers down Steele Rd. outside of Paris. This photo was used for a U.S. Army press release issued "for the papers of Sunday, May 2, 1943." The article released along with it apparently was intended to introduce the American public to Camp Tyson and its purpose. The Camp had been in operation for a year but was unknown to most Americans. Photo was taken at the intersection of Steele Rd. and Gate 3 Rd., where Lou and Bettye Carter now live. The photo is one of their prized possessions. Camp Tyson soldiers ran maneuvers through the hilly country roads around the camp on a daily basis. (Lou and Bettye Carter)

Children excitedly run after the barrage balloons during a parade in downtown Paris. Huge crowds lined the streets for the parade, which included soldiers, vehicles and balloons from Camp Tyson. (Virgil Wall)

African-American soldiers march with barrage balloons during a parade through downtown Paris. (Virgil Wall)

The interior of the former Tip Top Café, where African-American soldiers from Camp Tyson felt at home in Paris. The Tip Top Café was a fancy establishment at 418 W. Blythe St. where waitresses dressed in uniforms. In photo, several Military Police from the camp are visiting the Tip Top. Roland Atkinson waited tables there and said it was "a hopping place" The tall soldier standing directly under the light hanging from the ceiling was Lt. Peterson. The waitress at right in white uniform was Maude T. Sutton. (Roland Atkinson)

Camp Tyson with barrage balloon. (Charlie Allen)

An illustration of one the barrage balloons tied to a jeep during maneuvers at Camp Tyson. The illustration is one of many photos and memorabilia kept by former soldier Lowell Whiteside in his scrapbook.

The USOs for black and white soldiers were separate. The black USO was located in a large two-story building on Rison St., in Paris, which still stands. This photo shows the interior of the USO during an event. (Quinn Chapel AME Church)

The late James Wilson stands at right with fellow soldiers at Camp Tyson. Wilson was used a lot by the U.S. Army for press photos. (James Wilson)

"B" Battery getting ready for a battalion parade, May 1943. (Lowell Whitesides)

The morning call to arms at Camp Tyson. (James Wilson)

Soldiers on parade with one of the camp churches in the background. (James Wilson)

AS IF THEY WERE OURS

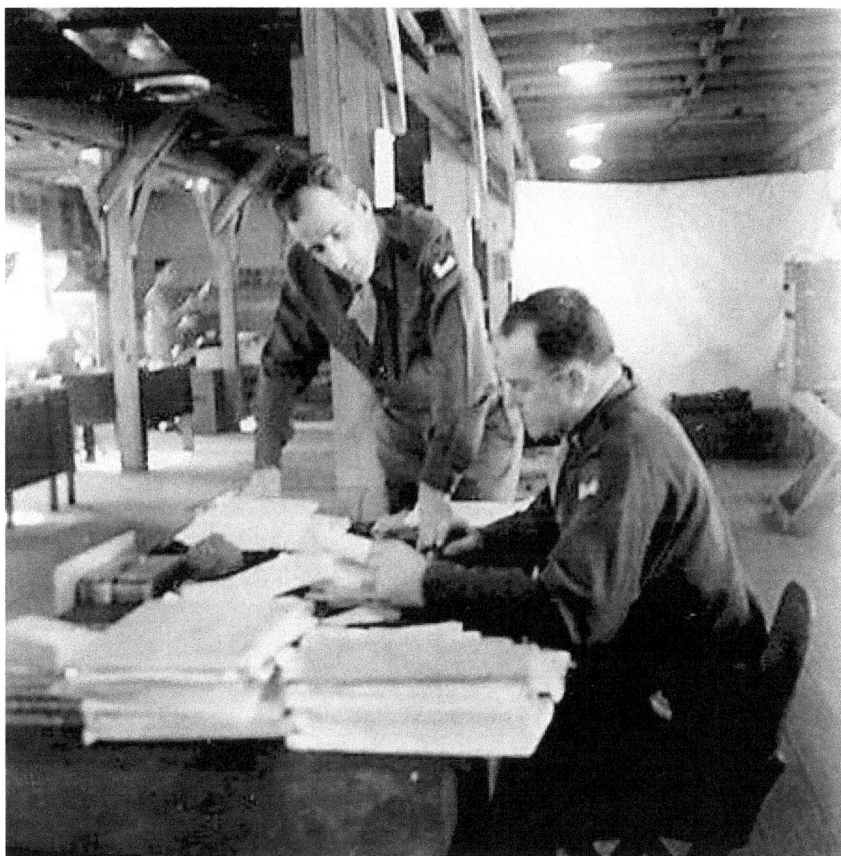

Camp Tyson administration doing paperwork. (State Library and Archives)

Camp Tyson Army Band Reunion.

A painting by the late Paul Soik, a former soldier at Camp Tyson. The painting is called "Haul 'Er Down" and it depicts soldiers deflating a barrage balloon after the day's training. Soik went on to become a nationally well-known graphic artist. The painting is shown being hung at the Paris-Henry Co. Heritage Center by Brig. Gen. Roland Parkhill, who preserved the painting, and former Heritage Center Director Norma Steele. (Author)

Joe Lankford points to one of many 'tie-downs' that are still on his property on Steele Rd. The tie-downs were used to secure the barrage balloons. Lankford's property used to be on the Camp Tyson grounds. Lankford is one of several property owners whose families were able to buy their land back after the Camp closed. (Author)

AS IF THEY WERE OURS

Joe Lankford points to one of several tie-downs that he has preserved on his property on Steele Rd. where the camp was located. The tie-downs were used to secure the barrage balloons. (Author)

The late Eddie Clericuzio of McKenzie enjoyed remembering his Camp Tyson days during an interview for the book. (Author)

The 90 foot tall hangar that was used to make and repair barrage balloons on the Camp Tyson grounds. The hangar is still there, used by the Spinks Co. which now owns the property. (Author)

Former Camp Tyson barracks are now used for offices by the Spinks Co. (Author)

The motor pool at Camp Tyson is now used as a garage for the Spinks Clay Co. (Author)

As If They Were Ours

With Camp Tyson being in the segregated South, the African-American and white soldiers had separate USO buildings. The USO for the black soldiers was located in this large, two-story building which still is located on Rison St. in Paris. (Author)

Camp Tyson utilized this crude, brick building as its furnace. The building is still located on the grounds where the camp was located. (Author)

The late Rebecca Goins of Paris kept a large suitcase full of documents which her father gathered on Camp Tyson history. (Author)

Former Spinks employee Chris Corley shows where the cemetery was located on Camp Tyson. The spot is now overgrown and bricks that had decorated the cemetery are now crumbling. (Author)

One of the singing groups consisting of African-American soldiers at Camp Tyson. The groups used to visit local black churches on Sundays, where they could feel at home, entertain and share fellowship. (Dorothy Cook)

The late Eddie Moody of Paris looking sharp in his uniform at Camp Tyson. Moody was a chauffeur for Camp Commander General Maynard and kept the memory of Camp Tyson alive by organizing camp reunions for years after the camp closed. (Tommy Moody)

As If They Were Ours

The late John Steele was hired by the federal government to mow the huge grounds that made up Camp Tyson. He used four teams of mules and was paid $1 an hour. His family still has his invoices. (Photo courtesy of Joe Lankford).

Pearl Routon smiles while driving her tractor, with barrage balloons afloat in the background. Pearl Routon is featured in Chapter 18. She and her family lived in the town of Routon, where Camp Tyson was located. She was postmistress of the town, operated the country store, the local restaurant and the cabins where many soldiers' families and camp workers stayed. With her close proximity to the camp, she was devoted to the local war effort. (Jill Routon Wilson)

The late Richard Carothers Sr. with his favorite horse Cow Lady. The Carothers family transformed the former Camp Tyson grounds into a working cattle and horse ranch and also located the Spinks Clay Company there. (Photo courtesy of Carothers' family scrapbook).

Shorty Hutcherson, right, and another cowboy at work at the horse ranch that was located at the former Camp Tyson grounds after the war. (Photo courtesy of Carothers' family scrapbook).

AS IF THEY WERE OURS

www.ingramcontent.com/pod-product-compliance
Lightning Source LLC
Chambersburg PA
CBHW051946090426
42741CB00008B/1298